ENVIRONMENTAL AND INDUSTRIAL CORROSION - PRACTICAL AND THEORETICAL ASPECTS

Edited by **Benjamin Valdez Salas**
and **Michael Schorr**

Environmental and Industrial Corrosion - Practical and Theoretical Aspects
http://dx.doi.org/10.5772/45617
Edited by Benjamin Valdez Salas and Michael Schorr

Contributors

Osama Abuzeid, Mohamed Abou Zour, Ahmed Ibrahim AlGoburi, Yahya Alzafin, Ryan Cottam, Agnieszka Kurc-Lisiecka, Hamid Reza Shahverdi, Hamoon Zohdi, Mohammad Emami, Nelly Cecilia Alba De Sánchez, Benjamin Valdez, Adina-Elena Segneanu, Ioan Grozescu, Nandina Vlatanescu, Ionel Balcu, Paula Sfarloaga

Published by InTech
Janeza Trdine 9, 51000 Rijeka, Croatia

Publishing Process Manager Ana Pantar
Technical Editor InTech DTP team
Cover InTech Design team

Additional hard copies can be obtained from orders@intechopen.com

Environmental and Industrial Corrosion - Practical and Theoretical Aspects, Edited by Benjamin Valdez Salas and Michael Schorr
p. cm.
ISBN 978-953-51-0877-1

For EU product safety concerns:
IN TECH d.o.o., Prolaz Marije Krucifikse Kozulić 3, 51000 Rijeka, Croatia, info@intechopen.com or visit our website at intechopen.com.

Meet the editors

Benjamin Valdez is the director of the Institute of Engineering of the University of Baja California, member of the Mexican Academy of Sciences and the National System of Researchers of Mexico. He was granted the B.S. degree in Chemical Engineering, the M.Sc. degree in Chemistry and the PhD. in Chemistry by the Autonomous University of Guadalajara. He has been Guest Editor for several special issues in journals on materials and member of the editorial boards of Corrosion Reviews, Corrosion Engineering Science and Technology and Revista Metalurgia. Dr. Valdez is the Coordinator of the National Corrosion Network. He is a full professor at the University of Baja California participating in activities on basic research, technological development and teaching postgraduate programs at the UABC and the Autonomous University of Campeche. His activities include research on the topics; corrosion and materials, electrochemical and industrial processes, chemical processing of agricultural and natural products, and consultancy in corrosion control in industrial plants and environments. He is the author of books, chapters in books and scientific articles on electrochemistry, biodeterioration and corrosion of materials and general chemistry. He is the head of the corrosion and materials department and the founder of the master and doctoral program in sciences and engineering for UABC. His professional career includes the collaboration with chemical, oil, water and energy industry of Mexico, and the preservation of the industrial infrastructure.

Michael Schorr is a Professor and Doctor Honoris Causa at the Institute of Engineering of the University of Baja California in Mexico and member of the Mexican Researchers System. He was granted the B.Sc. in chemistry and M.Sc. in materials engineering, both by the Technion in Israel and served as Editor of Corrosion Reviews, during the period 1986-2004. Professor Schorr has fifty years of experience in controlling environmental and industrial corrosion in Latin America, Europe, USA, South Africa and Israel. He authored and co-authored 240 scientific and technical publications on corrosion science, engineering and technology, as well as books in English, Spanish and Hebrew.

Contents

Preface

The environmental quality, world water scarcity and clean energy are established today as central disciplines in modern science, engineering and technology; they constitute an immense, critical problem for the whole humankind. They are already linked to the crucial issues of climate change, global warming and greenhouse emissions, all interrelated phenomena.

Nowadays, it is generally accepted that in the natural and industrial environments, corrosion and pollution are interrelated harmful processes since many water, air and soil pollutants accelerate corrosion, and corrosion products such as rust, oxides and salts, also pollute water bodies at ports, rivers, dams, canals and shipyards. Both are pernicious processes that impair the durability of the infrastructure assets, the industry efficiency and profitability and the quality of the environment. Therefore, it is essential to develop an to apply corrosion engineering control methods and techniques, in particular those being environment-friendly. This book will contribute to achieve this important mission.

In this crucial time of worldwide energy crisis and economic turmoil, cost effective corrosion control will extend the life of the infrastructure saving large expenses in materials, equipment and structures.

This volume is well-organized in three Parts: Industrial Corrosion, Corrosion Science and Technology and Corrosion of Biomaterials, which display the interaction of industrial and modern materials with different environments such as industrial, natural and physiological.

Chapters of the industrial corrosion section present a general overview for the corrosion processes at different industries and the combined effects of this phenomenon with mechanical aspects. Corrosion behavior of cold-deformed austenitic alloys and the stress corrosion cracking of ductile Ni-resist irons and stainless steels studies are good examples of failure caused by corrosion in processes for desalinating seawater where brine is a very corrosive environment rich in chlorides.

Other important mechanical factor that contributes to the corrosion processes is the wear by abrasion, erosion or cavitation. These effects are controlled by the metallic surface modification using different techniques for the application of special alloy coatings. Laser materials processing is one of the modern techniques described when it is utilized for corrosion performance improvement. The application of different layer and multilayer coatings for corrosion protection is discussed including welding mono and multilayer cladding on stainless steel surfaces. Finally, in the part on biomaterials corrosion, the corrosion characteristics of dental alloys, interesting pathological phenomena, is dealt by a comparison in vitro and in vivo, of the galvanic corrosion behavior of special alloys used as restoration, amalgam or prosthesis when bound together or coupled with dental implants.

This is a multi-authored book; the authors offer a wide spectrum of knowledge and experience. Since they are learned specialists in different areas and manifest different approaches and orientations, the book will be useful for both corrosion researchers and practitioners.

Dr. Benjamin Valdez Salas
Dr. Michael Schorr
Institute of Engineering
University of Baja California
Mexico

Section 1

Industrial Corrosion

Chapter 1

Laser Materials Processing for Improved Corrosion Performance

Ryan Cottam

Additional information is available at the end of the chapter

http://dx.doi.org/10.5772/52423

1. Introduction

Laser materials processing is a bourgeoning field of materials research. The power from a laser is used to heat, melt or ablate materials to change their character or topography. Their application to improve the corrosion performance of many types of metallic alloy and some composites can be broken down into four basic categories:

- laser melting, where the surface of the alloy is melted, which in turn changes the microstructure and in many cases improves corrosion,
- laser surface alloying, where a new material is introduced to surface during the laser melting process, which changes the microstructure and in many cases the corrosion performance.
- laser cladding, where a new more corrosion resistant material is added to the surface to improve the corrosion resistance,
- laser heating, where a solid state phase transformation is induced to change the microstructure to improve corrosion performance.

The field has been around for almost as long as laser material processing itself and it is the purpose of this chapter to discuss the development in these four areas, to give the reader an overview of the applications of laser material processing for improved corrosion performance.

2. Lasers and their operation

Laser as an acronym stands for 'Light Amplification by Stimulated Emission of Radiation'. Essentially laser are light amplifiers. The physics of how the light is amplified involves quantum mechanics, it is relatively complex and is comprehensively dealt with in [1]. The main difference between the different types of high powdered lasers is the medium used to generate the stimulate emission. These differences for the main types of high powered lasers will be dealt with below.

2.1. Main types of high powered lasers

2.1.1. CO2 lasers

Carbon dioxide laser were the first generation in industrially used high powered lasers, Figure 1. The amplification of the light is achieved through molecular vibration rather than electronic translations as in other lasers. The efficiency of CO2 lasers is around 10% which is quite low by today's standards and is the main reason for why this type laser are being used less and less. The wave length of CO2 lasers varies between 9.4 and 10.6μm which is quite large and because of this the light cannot be transferred by optical fibre and is typically achieved by mirrors, which has implications for applications. The cost of these lasers is relatively low and there are applications which they are suited and will continue to find application.

2.1.2. Neodynium – YAG lasers

This type of laser is finding less and less application but there are still many of them in use in research and industrial sectors. Abbreviated to Nd:YAG laser the amplification of light is achieved by triply ionised Nd as the lasant (a material that can be stimulated to produce laser light) and the crystal YAG (yttrium-aluminium-garnet) as the host. YAG is a complicated oxide with the chemical composition Y3Al5O12. The wavelength of this type of laser is 1.06 μm which is near the infrared spectrum. The light can be transport by optical fibre, which makes their application flexible and was responsible for their industrial uptake.

2.1.3. Diode lasers

Diode lasers are commonly used today and can be used in different configurations. Essentially as the name suggests diode lasers are didoes that have the ability to amplify light. Diodes are semiconductor materials for which there are many types and the wavelength produced range from 0.33 to 40μm. Of the known diode materials there are 20 that will lase. The most common ones are GaAs and AlxGa1-x. A variation on the diode laser principal is 'diode-pumped solid-state (DPSS) lasers '. This type of laser works by pumping a solid gain medium, like a ruby or a neodymium-doped YAG crystal, with a laser diode. This configuration of laser is compact and very efficient and is finding wide spread industrial use. The industrially used version of theses lasers can have the light transported by optical fibre which is important to their application.

Figure 1. POM DMD (direct metal deposition machine) featuring 6-axis rotating, CAD-CNC control using a 5kW CO2 Trumpf laser.

2.1.4. Fibre lasers

Fibre lasers are the latest technology in high powered lasers and can be banked together to produce powers in the range of 20-30kW. Fibre lasers use a doped optical fibre to amplify the light. The doping agents range from erbium, ytterbium, neodymium, dysprosium, praseodymium, and thulium. The high powers that can be generated with this technology are opening up new application for laser materials processing in the manufacturing sector and are currently the future of high powered lasers.

2.2. Laser optics

Apart from the CO2 laser, which use mirrors, the major types of high power lasers use optical fibre to deliver the light. Once near the region where the laser material processing is to occur, the light from the fibre needs to be collimated and then focused. Depending on the application the focal distance the distance can be varied by the choice of lenses for collimating and focusing. The profile of the light/beam is also dictated by the lens and can take the form of a Gaussian, top hat, bimodal (donut) or a line beam profile, Figure 2. For most applications a Gaussian beam is used, but top hat profile gives a sharp thermal profile in the material, which has advantages. Typically for laser materials processing applications the focal point is about 1mm in diameter and has a top hat profile. Above focus the beam profile changes from top hat to bimodal, which increasing radius and below focus the beam is Gaussian with increasing diameter, Figure 2.

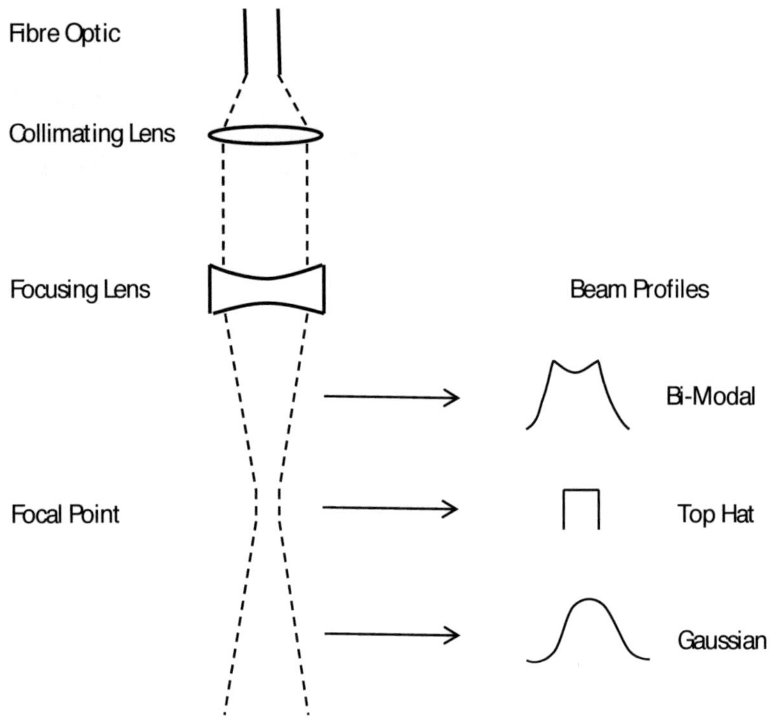

Figure 2. Laser optical configuration and resulting beam profile.

2.3. Controlling the movement of the laser

In the early days of laser materials processing the laser and optics were fixed to a CNC table and basic movement in X,Y and Z direction were possible. Nowadays robots are readily available and affordable, laser can be couple to the head of a robot allowing five-axis movement. Also computer programs and CAD drawings can be used to develop paths for laser movement, which can allow complex geometries to be processed. This development has been crucial to the application of laser materials processing technology and has applications for treating manufactured component that required improved corrosion resistance. A sixth degree of movement can also be obtained through the use of a rotating chuck that can grip the component to be processed and is common on many commercial laser cladding systems.

http://dx.doi.org/10.5772/52423

2.4. Materials preparation

Not all of the energy of the laser is transferred into heat in the substrate material. In fact the condition of the substrate plays a big role in how much energy is absorbed. In the as machined state metallic materials are highly reflective and between 0.1 - 0.3 of the energy is absorbed by the laser. The range is because some metals absorb the laser energy better than others. Coatings can be applied to the substrate to improve the absorption and in many cases the absorption can be as high as 0.9 of the energy of the beam. Sand or grit blasting reduces the reflectiveness of metallic materials and the rough surface also improves absorption and absorption values as high as 0.5 can be achieved. It should also be noted that the angle of incidence also affects the absorption of the laser. This is known as the Brewster effect and needs to be considered when processing complex geometries.

3. Laser processing techniques for improving surface corrosion performance of alloys

3.1. Laser surface melting

Laser surface melting (LSM) is performed by heating a metallic substrate using a laser with high enough power to create a melt pool. The melt pool travels with laser movement, which when coupled with a raster pattern, an area of material can be melted. The laser power level required to melt the material is dependent on the thermal diffusivity, conductivity and melting point of the substrate material as well as the rate at which the laser is being traversed. Because only a small portion of the substrate material is being melted the cooling rate is high and ranges from 103 – 106 °C/s depending on the thermo-physical properties of the substrate material and the traversing rate of the laser. This can result in new types of microstructures formed, which are typically more homogeneous and exhibit improved corrosion performance. The geometry of the melt pool during LSM is dependent on the power density and hence laser traversing speed. With increasing laser traversing speed the melt pool geometry changes from hemispherical to flat-bottomed with increase of traversing speed as thermal diffusion becomes limited. The flat-bottomed shape is the most desired for LSM.

3.1.1. Improving the pitting potential of alloys

By far most of the research into LSM for improved corrosion performance has been conducted on improving the pitting potential of commercial alloys. The general tendency of LSM to increase the homogeneity of the surface of a treated alloy is the reason for its application to increase pitting potentials. Low pitting potentials are associated with galvanic couples that can exist between second phase particles and the matrix which lower the potential for corrosion to occur and cause localised corrosion at the interface between the particles and the matrix, forming pits [2]. By dissolving the particles this mechanism is eliminated and the pitting potential is increased. The increase in the pitting potential is dependent of the potential of the galvanic couple of the particle and the matrix and the effect of the dissolved alloying element of the oxide layer formed and its ability to form a passivation film. While for many alloys the pitting potential is improved in some cases the pitting potential can be reduced if the alloying elements are not completely dissolved and the increase in the number of grain boundaries and fine second phase particles increases the pitting potential by increase the number of regions for pitting to occur. An example of pitting potential diagram showing how the pitting potential is determined is shown in Figure 3.

LSM has been trialled on a range of stainless steels to improve the pitting potential. The sensitisation of stainless steels during welding and other thermo processes is an issue for manufacturing stainless steels that will be used in corrosive environments. Essentially the thermal processes cause the formation of carbides with chromium, which locally depletes chromium available to form a passivation film and hence reduces the corrosion performance. LSM has proven to dissolve these carbides and restore the chromium levels to form the passivation film [3]. This can be applied to the welds of stainless steels to restore corrosion performance.

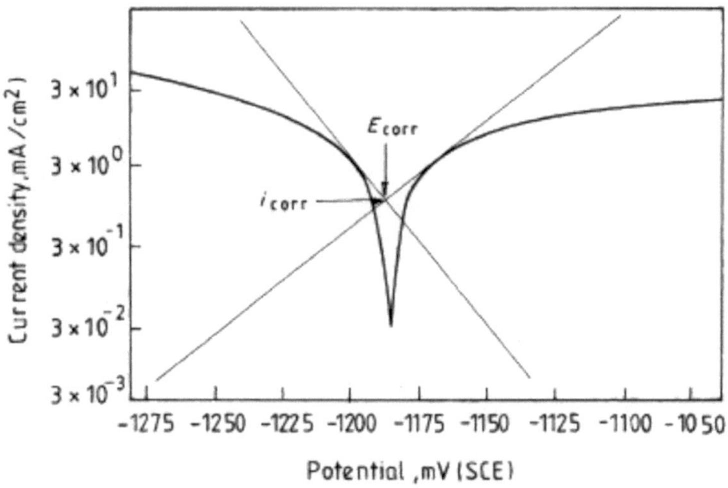

Figure 3. Variation of current density with potential indicating the pitting potential and current, taken from [3].

It has also been shown that LSM can change the surface chemistry of AISI 304 stainless steel to improve the pitting potential [4]. It was found that Cr_2O_3, Fe_2O_3, FeO and MnO_2 were detected on the surface only after LSM and was used to explain the higher pitting potential of the LSM treatment. LSM of 3CR12 stainless steel revealed that pitting potential was also enhanced by redistribution of non-intermetallic inclusions and the formation of a homogenous ferritic microstructure. Kwok et al [5] has performed the most comprehensive investigation into the pitting behaviour of AISI 440C martensitic stainless steels. It was found that larger carbides are more harmful to corrosion resistance than fine carbides due to greater inhomogeneity in the former case. It was also found that the formation of retained austenite was favourable for increasing the pitting potential. The laser scanning speed also influenced the pitting potential and was attributed to the effect of power density. In summary a range of stainless steels show an increase in pitting potential due to LSM and can be attribute to enhanced surface chemistry and an increase in the homogeneity of the microstructure.

Aluminium alloys have shown a mixed response in pitting potential due to LSM. 2xxx aluminium alloys have received the most attention. While 2014 has shown an improvement in pitting potential due to LSM it was found that the most significant result was a change in the morphology of the pits to be shallower due to both the distribution of copper and the refinement of the microstructure [4]. 2024 shows no significant change in the pitting potential due to LSM [5], it was found that the sites for corrosion shifted from the second-phase precipitates to the α-aluminium dendrites and with the elimination of inter-granular corrosion. Al-9wt% Si casting was also subjected to LSM and again the pitting potential was not improved [6]. This was attributed to an increase in the silicon aluminium boundaries caused by the refinement of the microstructure by LSM because these boundaries are more susceptible to the corrosion action. Aluminium alloy 6013 shows improved pitting resistance due to LSM brought about by a more homogeneous microstructure which increased the initial resistance to pitting [7].

The poor corrosion properties of magnesium alloys have resulted in many investigations using LSM to improve the pitting potential. AZ91 has received the most attention [8-10] where it was found that LSM dissolves the β-Mg17Al12 precipitate present in the as cast alloy which was responsible for micro galvanic corrosion and the formation of α solid solution of aluminium of up to 10wt%, which provides passive properties to the melted surface. ACM720 has also been investigated [11]and it was found that the rapid solidification of LSM supressed of the formation of Al2Ca phase which is present in the cast alloy in the grain boundaries. This change coupled with the extended solubility of aluminium in the microstructure increased the pitting potential. Rare earth containing alloy MEZ showed improved corrosion resistance [3]. It was suggested that the decrease in the anode cathode area due to the refinement of the microstructure coupled with the extended solubility of the rare earths increase the pitting corrosion resistance.

Ti-6Al-4V has also been subjected to LSM and it was found that the refinement of the microstructure improved the pitting potential in both Hank's solution and 3%NaCl solution [12, 13]. Alloy 600 also responded well to LSM and due to its high chromium content and forms an excellent passivation film during testing [14]. High speed tools steels M2, ASP23 and ASP30 were subjected to LSM and it was found that the pitting corrosion potential increased. This increase was attributed to the dissociating and refinement of large carbides and the increase of the passivation alloying elements in the ultrafine solid solution of austenite and martensite. Metal matrix composites have also responded well to LSM [15-17]. The corrosion behaviour of metal matrix composites is due to heterogeneity of their structure. LSM increase the homogeneity by dissolving the reinforcement phase.

3.1.2. Cavitation erosion and corrosion

Cavitation erosion corrosion occurs due to the formation and collapse of bubbles in fluid near the metallic surface. When a metal surface is subjected to high speed fluid flow sudden changes of liquid speed cause changes in the local vapour pressure which induces the formation and collapse of bubbles. The collapse of the bubbles generates shockwaves, which impact on the metal surface and leads to fatigue damage. LSM has been used on several types of stainless steels to improve the resistance to this form of corrosion [18-20]. The response of different alloys varies for example S31603 shows improved performance while S32760 and S30400 showed a decrease in the cavitation erosion corrosion performance. The decrease in performance for the two alloys was attributed to increased levels of twining in the two alloys. For martensitic stainless steel S42000 the cavitation erosion corrosion performance was improved through increased levels of retained austenite, which for this alloy is highly transformable. The stresses of the cavitation process are absorbed more readily due to the stress inducing a transformation of the retained austenite, hence increasing the resistance of the material to this form of corrosion.

3.1.3. Stress corrosion cracking

Stress corrosion cracking (SCC) is a form of cracking where an applied stress or residual stress in a component in combination with a corrosive environment acts to accelerate crack growth of the component. LSM has been conducted on AA 7075 in two separate studies [21, 22] to counteract this effect. Both studies found that the homogeneous refined microstructure produced by LSM delayed the onset of SCC when compared to the untreated material. Electrochemical impedance measurements showed that the film resistance was higher than that of the untreated material and provided an effective barrier to corrosion attack. It was also found that a nitrogen atmosphere during processing forms AlN phase which improves the corrosion resistance of the film further.

3.2. Laser surface alloying

Laser surface alloying (LSA) is where a preplaced powder is melted with the substrate to form a layer with a combined composition. The high cooling rates associated with the process novel microstructures are formed, which can exhibit improved corrosion performance. Like for LSM, process parameters are dependent on the thermo-physical properties of the substrate and alloying material as well as the processing speed. The level of dilution plays a big role in the composition of the melted layer, which in turn can have a dramatic effect on the microstructure produced. There are several techniques for preplacing the alloying element(s), the powder can be combined with a polymer to form a paste, which is then smeared onto the surface to be alloyed. Flame spray is another method to preplace the powder as is hot dipping.

One avenue to improve the corrosion performance using LSA is to alloy with an element, which forms a strong passivation film. For example, alloying AA7175 with chromium to improve its crevice corrosion performance [23], the alloying of ductile iron with copper [24] and the alloying of copper with titanium[25]. Another method is to alloy to form an intermetallic with alloying elements that exhibit good corrosion performance such as the alloying of steels with aluminium to form iron aluminides [26]. The area that has received the most attention is alloying to improve the cavitation corrosion properties. LSA is ideal for this application because not only can the corrosion resistance of the surface be improved by alloying but the hardness of the surface can be improved as well, which are both properties required of a cavitation corrosion resistant material. Aluminized steel prepared by LSA showed a 17 fold increase in the cavitation corrosion resistance [27]. The high hardness of the intermetallic layer and the shift in the pitting potential to a more noble level was used to explain the dramatic increase. Ni-Cr-Si-B alloy was LSAed with brass for improved cavitation corrosion[28]. Variations in the power density during processing influence the cavitation corrosion behaviour and it was found that it was a compromise between cavitation erosion performance and corrosion, the harder layers exhibited improve cavitation corrosion and while the corrosion is improved by a more homogeneous microstructure. A substantial investigation into LSA of stainless steel has been conducted by Kwok et al [29]. UNS S31603 was alloyed with Co, Ni, Mn, C, Cr, Mo, Si. As would be expected a vast variety of microstructures were produced and were interpreted with reference to the appropriate phase diagrams. Essentially the layers alloyed with Co, Ni, Mn, C contained austenite as the main phase while alloying with Cr and Mo resulted in ferrite. While the corrosion was improved in some cases, there was no underlying trend between the different alloying elements. This highlights the complexity of the microstructural formation in LSA and its dependence on process parameters.

3.3. Laser cladding

Laser cladding (LC) is a process where a new layer is created on the surface of a metal/alloy. A traversing laser is used to heat the substrate and form a molten pool. Into the pool either a wire or a powder is blown of a desired composition, which then melts and then quickly solidifies when past the laser beam, forming a new layer. The spot size of the laser, the traversing speed, the powder feed rate and power of the laser influences the resulting clad layer. By cladding materials that exhibit good corrosion performance the corrosion performance of a component can be increased.

This field of research has received limited attention. The cladding of stainless steels on mild steel has been investigated and like for laser surface melting the high cooling rates produce a fine microstructure, which coupled with the excellent corrosion properties of stainless steel the corrosion performance of the surface of the mild steel was improved [30]. The effect of La2O3 on ferritic steel was investigated [31] an proved to increase the corrosion resistance of the substrate by increasing the passivation effects of the film during testing. Nickel silicide as a clad layer have also been investigated [32]. This particular type of intermetallic displayed excellent corrosion response, which was attributed to the large amount of silicon which promotes the formation of highly dense and tightly adherent passive thin film when contacting with corrosive media.

3.4. Laser heat treatment

Laser heat treatment (LHT) is where the laser heats the substrate without melting the top surface layer. The heating then allows solid-state phase transformation(s) to occur which when coupled with the high cooling rates of laser materials processing, new microstructures can be formed. This is a new field of laser material processing for improved corrosion performance and has the advantage over LSM in that the residual stress after processing is lower.

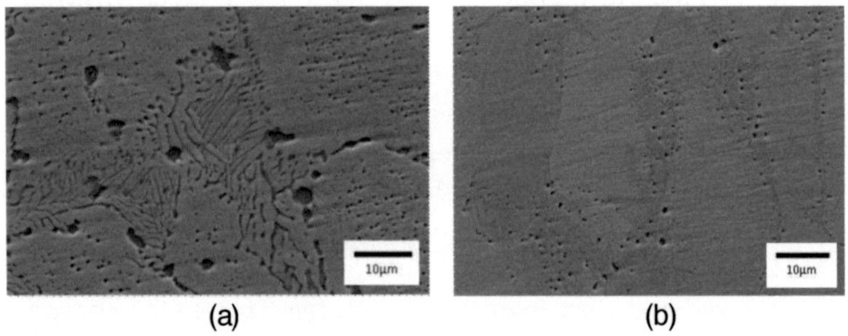

Figure 4. Nickel-aluminium bronze microstructure before and after laser heat treatment processing; a – Cast microstructure showing precipitates and κ_{III} lamella; b – laser heat treated microstructure showing that the κ_{IV} and κ_{III} lamella was dissolved.

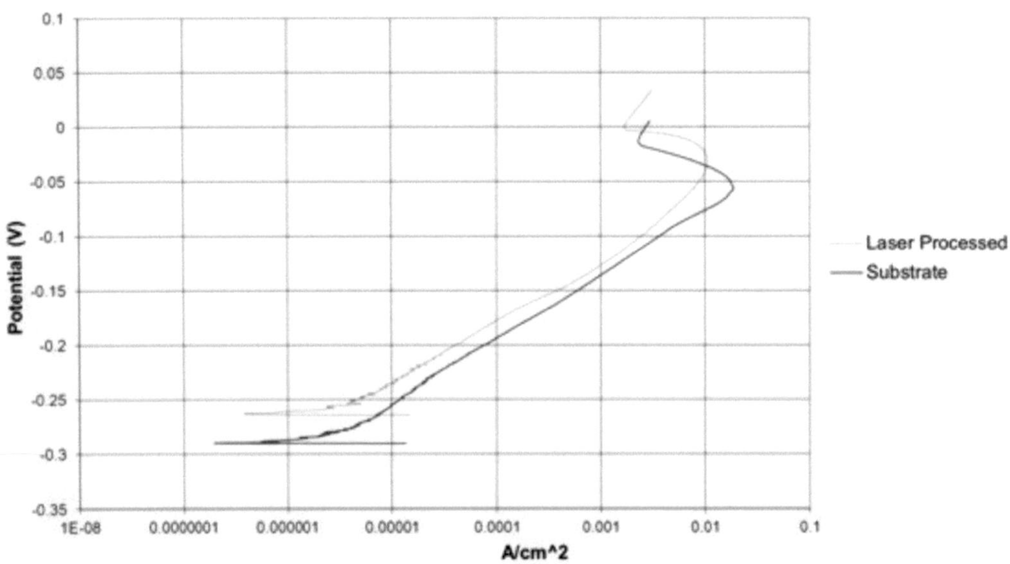

Figure 5. Linear polarisation curves for as cast substrate and laser processed nickel-aluminium bronze.

An example of this is laser transformation hardening of steels. The heat from the laser causes the ferrite and pearlite to transform to austenite and then the rapid cooling provided by conduction to the substrate produces hard martensite. Typically this technique is uses to produce a wear resistant layer, however it has been applied stainless steel for improved cavitation corrosion resistance [20]. It was found that the ratio of austenite to martensite was better than for LSM. Another example of this is the laser heat treatment of nickel-aluminium bronze (NAB). LC of NAB produces the Widmanstatten morphology microstructure, which has poor corrosion properties [33]. Whereas the LHT of NAB produces a microstructure that is free form precipitates, Figure 4, which is responsible for the reduced corrosion performance and is free from the Widmanstatten morphology microstructure [34]. This new type of microstructure exhibited improved corrosion properties as shown by the linear polarisation results, Figure 5.

3.5. Hybrid coatings

Laser melting or heating can also be used in conjunction with other coating techniques. Thermal spray which covers a range of processing techniques applies coatings to metallic substrates by heating a powder and propelling into a substrate. Each individual particle forms a splat on the substrate and the accumulation of many splats result in a coating. While these coating have found many applications including thermal barrier coatings and wear resistant coatings they have flaws. They are that the coatings have poor bond strength between the coating and the substrate, and high porosity. Laser melting offer the opportunity removes these two issues and so hybrid coating applications have been investigated.

In relation to developing coatings for improved corrosion resistance the hybrid thermal spray laser melting have found applications with Ni coated WC coatings [35, 36], thermal barrier coatings [37, 38] super alloy WC coatings [39] Al-SI alloy coatings on AZ91D magnesium alloy [40] and NiCrBSi alloys [41, 42]. For all these applications the laser melting consolidated and homogenised the coating which improved the corrosion performance. In some cases if the wrong laser parameters were used cracking of the coating occurred mitigating the effect of reducing the porosity and should be considered when developing this type of coating.

The second group of hybrid coatings and one that has received less attention is electroplating followed by laser melting. Electroplated coatings like thermal spray coatings have porosity and therefore their corrosion performance is reduced. Laser melting of the coatings consolidates the porosity and improves their performance. Examples of this are Ni-W-P coatings [43] and gold coatings [44].

3.6. Superhydrophobic surfaces produced by femtosecond lasers

Superhydrophobic surface are a biomimetic surface structure where nano scale features on their surface cause a hydrophobic response [45]. These surface can be produced on metallic materials using femtosecond laser technology (pulsed laser ablation) [46]. The ability to repel liquids, which cause the corrosive action and the nano-scale of their structure could combine to enhance corrosion performance including bio-corrosion, where the microbes are responsible for the corrosion process. These surface have successfully produced on stainless steel [47]. While there corrosion performance has not been evaluated, this may be the way of the future for specific areas of corrosion prevention.

4. Conclusion

High powered lasers have found many applications in improving the corrosion performance of metallic materials and some composite materials. The high cooling rates associated with laser processing promotes the formation of more homogeneous microstructures, which in turn improves the corrosion performance. As has been shown a wide range of alloy systems can utilise this technology and potentially there are other systems for which the techniques described could be applied. Laser alloying is an area where more work could occur in the further by carefully selecting alloying elements, with reference to phase diagrams, to optimise the effect of alloying. Superhydrophobic surfaces are also an area that could be explored as well. Laser material processing is a growing field of research and as it grows its application to corrosion science and prevention will also increase.

Acknowledgements

This work has been conducted by funding from the Defence Materials Technology Centre (DMTC), program 2, project 2.2 "Surface Processing Technologies for Repair and Improved Performance of Submarine and Surface Ship Components". The images in figures 4 and 5 were supplied from DMTC project 2.2.

Author details

Ryan Cottam[1,2]

1 Industrial Laser Applications Laboratory, IRIS, Faculty of Engineering and Industrial Sciences, Swinburne University of Technology, Victoria, Australia

2 Defence Materials Technology Centre (DMTC), Hawthorn, Victoria, Australia

References

[1] Migliore, L., Laser Materials Processing1996: Marcel Dekker. 319.

[2] Liu, Z., P.H. Chong, P. Skeldon, P.A. Hilton, S. J.T., and B. Quayle, Fundamental understanding of the corrosion performance of laser-melted metallic alloys. Surface and Coatings Technology, 2006; 200 5514-5525.

[3] Majumdar, J.D., R. Galun, B.L. Mordike, and I. Manna, Effect of laser surface melting on corrosion and wear resistance of a commercial magnesium alloy. Materials Science and Engineering A, 2003; 361 119-129.

[4] Chong, P.H. and Z. Liu, Large area laser surface treatment of aluminium alloys for pitting corrosion protection. Applied Surface Science, 2003. 208-209: p. 399-404.

[5] Li, R., M.G.S. Ferreira, A. Almeida, R. Vilar, K.G. Watkins, M.A. McMahon, and W.M. Steen, Localisation of laser surface melted 2024-T351 aluminium alloy. Surface and Coatings Technology, 1996; 81 290-296.

[6] Osorio, W.R., N. Cheung, J.E. Spinelli, K.S. Cruz, and A. Garcia, Microstrucuarl modification by laser surface remelting and its effect on the corrosion resistance for an Al-9 wt%Si casting alloy. Applied Surface Science, 2008; 254 2763-2770.

[7] Xu, W.L., T.M. Yue, H.C. Man, and C.P. Chan, Laser surface melting of aluminium alloy 6013 for improving pitting corrosion fatigue resistance. Surface and Coatings Technology, 2006; 200 5077-5086.

[8] Gao, Y., C.M. Wang, M. Yao, and H. Liu, Corrosion behviour of laser melted AZ91HP magnesium alloy. Materials and Corrosion, 2007; 58 463-466.

[9] Guan, Y.C., W. Zhou, and H.Y. Zheng, Effect of laser surface melting on corrosion behaviour of AZ91D Mg alloy in simulated-modified body fluid. Journal of Applied Eletcrochemistry, 2009; 39 1457-1464.

[10] Coy, A.E., F. Viejo, F.J. Garcia-Garcia, Z. Liu, P. Skeldon, and G.E. Thompson, Effect of eximer laser surface melting on the microstructure and corrosion performance of the die cast AZ91D magnesium alloy. Corrosion Science, 2010; 52 387-397.

[11] Mondal, A.K., S. Kumar, C. Blawert, and N.B. Dahotre, Effect of laser surface treatment on corrosion and wear resistance of ACM720 Mg alloy. Surface and Coatings Technology, 2008; 202 3187-3198.

[12] Yue, T.M., T.M. Cheung, and H.C. Man, The effects of laser surface treatment on the corrosion properties of Ti-6Al-4V alloy in Hank's solution. Journal of Materials Science Letters, 2000; 19 205-208.

[13] Sun, Z., I. Annergren, D. Pan, and T.A. Mai, Effect of laser surface remelting on the corrosion behaviour of commercially pure titanium sheet. Materials Science and Engineering A, 2003; 345 293-300.

[14] Shin, J.K., J.H. Suh, J.S. Kim, and S.-J.L. Kang, Effect of laser surface modification on the corrosive resistance of Alloy 600. Surface and Coatings Technology, 1998; 107 94-100.

[15] Yue, T.M., Y.X. Wu, and H.C. Man, Improvement in the corrosion resistance of aluminium 2009/SiCw composite by Nd:YAG laser surface treatment. Journal of Materials Science Letters, 1999; 18 173-175.

[16] Yue, T.M., Y.X. Wu, and H.C. Man, Laser surface treatment of aluminium 6013/SiCp composite for corrosion resistance enhancement. Surface and Coatings Technology, 1999; 114 13-18.

[17] Hu, J.D., P.L. Wu, L.C. Kong, and G. Liu, The effect of YAG laser surface treatment on corrosion resistance of Al18B4O33 w/2024Al composite. Materials Letters, 2007; 61 5181-5183.

[18] Kwok, C.T., H.C. Man, and F.T. Cheng, Cavitation erosion and pitting corrosion of laser surface melted stainless steels. Surface and Coatings Technology, 1998; 99 295-304.

[19] Kwok, C.T., H.C. Man, and F.T. Cheng, Cavitation erosion and pitting corrosion behaviour of laser surface-melted martensitic stainless steel UNS S42000. Surface and Coatings Technology, 2000; 126 238-255.

[20] Lo, K.H., F.T. Cheng, C.T. Kwok, and H.C. Man, Effects of laser treatments on cavitation erosion and corrosion of AISI 440C martensitic stainless steel. Materials Letters, 2003; 58 88-93.

[21] Yue, T.M., C.F. Dong, L.J. Yan, and H.C. Man, The effect of laser surface treatment on stress corrosion cracking behaviour of 7075 aluminium alloy. Materials Letters, 2004; 58 630-635.

[22] Yue, T.M., L.J. Yan, and C.P. Chan, Stress corrosion cracking behaviour of Nd:YAG laser-treated aluminium alloy 7075. Applied Surface Science, 2006; 252 5026-5034.

[23] Ferreira, M.G.S., R. Li, and R. Vilar, Avoiding crevice corrosion by laser surface treatment. Corrosion Science, 1996; 38 2091-2094.

[24] Zeng, D., C. Xie, Q. Hu, and K.C. Yung, Corrosion resistance enhancement of Ni-resist ductile iron by laser surface alloying Scripta Materialia, 2001; 2001 651-657.

[25] Wong, P.K., C.T. Kwok, H.C. Man, and F.T. Cheng, Corrosion behaviour of laser-alloyed copper with titanium fabricated by high power diode laser. Corrosion Science, 2012; 57 228-240.

[26] Abdolahi, B., H.R. Shaverdi, M.J. Torkamany, and M. Emami, Improvement of the corrosion behaviour of low carbon steel by laser surface alloying. Applied Surface Science, 2011; 257 9921-9924.

[27] Kwok, C.T., F.T. Cheng, and H.C. Man, Cavitation erosion and corrosion behaviours of laser-aluminized mild steel. Surface and Coatings Technology, 2006; 200 3544-3552.

[28] Tam, K.F., F.T. Cheng, and H.C. Man, Enhancement of cavitation erosion and corrosion resistance of brass by laser surface alloying with Ni-Cr-Si-B. Surface and Coatings Technology, 2002; 149 36-44.

[29] Kwok, C.T., F.T. Cheng, and H.C. Man, Laser surface modification of UNS S31603 stainless steel. Part I: microstructures and corrosion characteristics. Materials Science and Engineering A, 2000; 290 55-73.

[30] Li, R., M.G.S. Ferreira, M. Anjos, and R. Vilar, Localized corrosion performance of laser surface cladded UNS S4470 superferritic stainless steel on mild steel. Surface and Coatings Technology, 1996; 88 96-102.

[31] Zhao, G.M. and K.L. Wang, Effect of La2O3 on corrosion resistance of laser clad ferrite-based alloy coatings. Corrosion Science, 2006; 48 273-284.

[32] Cai, L.X., H.M. Wang, and C.M. Wang, Corrosion resistance of laser clad Cr-alloyed Ni2Si/NiSi intermetallic coatings. Surface and Coatings Technology, 2004; 182 294-299.

[33] Hyatt, C.V., K.H. Magee, and T. Betancourt, The Effect of heat Input on the Microstructure and Properties of Nickel Aluminium Bronze Laser Clad with a Consumable of Composition Cu-9.0Al-4.6Ni-3.9Fe-1.3Mn. Metallurgical and Materials Transactions A, 1998; 29A 1677-1690.

[34] Cottam, R. and M. Brandt, Development of a Processing Window for the Transformation Hardening of Nickel-Aluminium-Bronze. Materials Science Forum, 2010; 654-656 1916-1919.

[35] Xie, G., J. Zhang, Y. Lu, Z. He, B. Hu, D. Zhang, K. Wnag, and P. Lin, Influence of laser treatment on the corrosion properties of plasma-sprayed Ni-coated WC coatings. Applied Surface Science, 2007; 253 9198-9202.

[36] Guozhi, X., Z. Jingxian, L. Yijun, W. Keyu, M. Xiangyin, and L. Pinghua, Effect of laser remelting on corrosion behaviour of plasma-sprayed Ni-coated WC coatings. Materials Science and Engineering A, 2007; 460-461 351-356.

[37] Tsai, P.-C. and C.S. Hsu, High temperature corrosion resistance and microstructural evaluation of laser-glazed plasma-sprayed zirconia/MCrAlY thermal barrier coatings. Surface and Coatings Technology, 2004 183 29-34.

[38] Tsai, P.C., J.H. Lee, and C.S. Hsu, Hot corrosion behaviour of laser-glazed plasma-sprayed yttria-stabilized zirconia thermal barrier coatings in the presence of V2O5. Surface and Coatings Technology, 2007; 201 5143-5147.

[39] Liu, Z., J. Cabrero, S. Niang, and Z.Y. Al-Taha, Improving corrosion and wear performance of HVOF-sprayed Inconel 625 and WC-Incionel 625 coatings by high power diode laser treatments. Surface and Coatings Technology, 2007; 201 7149-7158.

[40] Qian, M., D. Li, S.B. Liu, and S.L. Gong, Corrosion performance of laser-remleted Al-Si coating on magnesium alloy AZ91D. Corrosion Science, 2010; 52 3554-3560.

[41] Navas, C., R. Vijande, J.M. Cuetos, M.R. Fernandez, and J. de Damborenea, Corrosion behviour of NiCrBSi plasma-sprayed coatings partially melted with laser. Surface and Coatings Technology, 2006; 201 776-785.

[42] Serres, N., F. Hlawka, S. Costil, C. Langlade, and F. Machi, Corrosion properties of in situ laser remelted NiCrSi coatings comparison with hard chromium coatings. Journal of Materials Processing Technology, 2011; 211 133-140.

[43] Liu, H., F. Viejo, R.X. Guo, S. Glenday, and Z. Liu, Microstructure and corrosion performance of laser-annealed electrolyss Ni-W-P coatings. Surface and Coatings Technology, 2010; 204 1549-1555.

[44] Georges, C., H. Sanchez, N. Semmar, C. Boulmer-Leborgne, C. Perrin, and D. Simon, Laser treatment for corrosion prevention of electrical contact gold coating. Applied Surface Science, 2002; 186 117-123.

[45] Guo, Z., W. Liu, and B.-L. Su, Superhydrophobic surfaces: From natural to biomimetic funcitonal. Journal of Colloid and Interface Science, 2011; 353 335-355.

[46] Kietzig, A.-M., M.N. Mirvakili, S. Kamal, P. Englezos, and S.G. Hatzikiriakos, Laser-Pattered Super-Hydrophobic Pure Metallic Subtrates: Cassie to Wenzel wetting Transitions. Journal of Adhesion Science and Technology, 2011; 25 2789-2809.

[47] Wu, B., M. Zhou, J. Li, X. Ye, and L. Cai, Superhydrophobic surfaces fabricated by microstructuring of stainless steel using femtosecond laser. Applied Surface Science, 2009; 256 61-66.

Chapter 2

Corrosion Control in Industry

B. Valdez, M. Schorr, R. Zlatev, M. Carrillo,
M. Stoytcheva, L. Alvarez, A. Eliezer and N. Rosas

Additional information is available at the end of the chapter

http://dx.doi.org/10.5772/51987

1. Introduction

The economic development of any region, state or country, depends not only on its natural resources and productive activities, but also on the infrastructure that account for the exploitation, processing and marketing of goods. Irrigation systems, roads, bridges, airports, maritime, land and air transport, school buildings, offices and housing, industrial installations are affected by corrosion and therefore susceptible to deterioration and degradation processes.

Corrosion is a worldwide crucial problem that strongly affects natural and industrial environments. Today, it is generally accepted that corrosion and pollution are interrelated harmful processes since many pollutants accelerate corrosion and corrosion products such as rust, also pollute water bodies. Both are pernicious processes that impair the quality of the environment, the efficiency of the industry and the durability of the infrastructure assets. Therefore, it is essential to develop and apply corrosion engineering control methods and techniques.

Other critical problems, that impact on infrastructure and industry are climate change, global warming and greenhouse emissions, all interrelated phenomena.

This chapter presents important aspects of corrosion in industrial infrastructure, its causes, impacts, control, protection and prevention methods.

2. Materials in industry

Metallic materials play a key role in the development of a country and its sustained growth in the context of the global economy. Table 1 shows a classification and the properties of different types of materials used in the industry. During the course of the metal production it undergoes various types of processes: mining of minerals, manufacturing and application and generation of gases, liquids or solids that are released into the environment. In the industrial development, production and use of materials in general, economic cycles are due to take effect that influence the environment (Raichev et al., 2010). The selection of a predominant group of materials depends on the particular industries; they determine to a greater or lesser extent the pattern of consumption of a given product, inducing the market to adapt itself to this new reality. The materials industry follows two general strategies: research the materials and the available technology recommended for their. Recycled materials typically require less capital and energy consumption, but need more manpower, for primary processing. Also, the costs of pollution control are lower than those required for primary processing of minerals. Recycling becomes more intense, as economies tend to be more sophisticated, since viable quantities of recycled material must be available for reuse (Garcia, R., et al, 2012, Lopez, G. 2011, Schorr, M., 2010).

Material	Main properties	Uses
Metals and alloys (carbon and stainless steels, non ferrous alloys)	Mechanical resistance hardness	Cars, aircraft, tanks, infrastructure reinforcement.
Plastics (Synthetic polymers, rubbers)	Low density and corrosion resistance	Process components, tubes, vessels, coatings, paints.
Ceramics (Metallic carbides, silica, glass, alumina)	High hardness, high temperature and corrosion resistance	Cutting tools, motor components, refractory bricks, ovens, etc.
Composites (glass and carbon fibers reinforced plastics, plastic matrixes reinforced with metallic particles)	Light weight, high strength and hardness.	Car bodies, aircraft components, vessels, construction.

Table 1. Materials in industry: Types, main properties and uses.

In the production of a material waste is generated: for example, parts of material that was left aside, through the production steps. There are called effluent, which consist of waste that comes from the processes linked to the technology involved in each step of production, although not necessarily with the main material. Industrial processes for the recovery of ore from the mine to produce a metal, are related to technological development and therefore varies from one country to another, including regulatory laws, financial aspects etc.. Therefore, the environmental impacts vary widely. A low grade or poor quality of the ore, with low metal content, increase the cost of recovery, requiring large amounts of mineral raw material and energy invested for the recovery of small amounts of metal. Also important is the feasibility of the mineral that can be worked out e.g., the cost of physical removal of rock, accessibility to the mines, thickness and regularity of the ore zone, and its hardness. Figure 1, shows the material cycle, which involves processes from raw material, extraction from natural sources, processing

http://dx.doi.org/10.5772/51987

and conversion into industrial materials, their processing and application, the deterioration rate effects, its mechanical properties, environmental behavior, corrosion, disposal and possible recovery of some of these through the use of recycling methods.

There are many examples of recovery of metals, which could help to describe step by step the various interactions with the environment itself. A mineral submitted to a production process will impact the environment, during four steps: extraction, processing, fabrication and manufacturing, of goods as seen in the cycle of materials. (Figure 1)

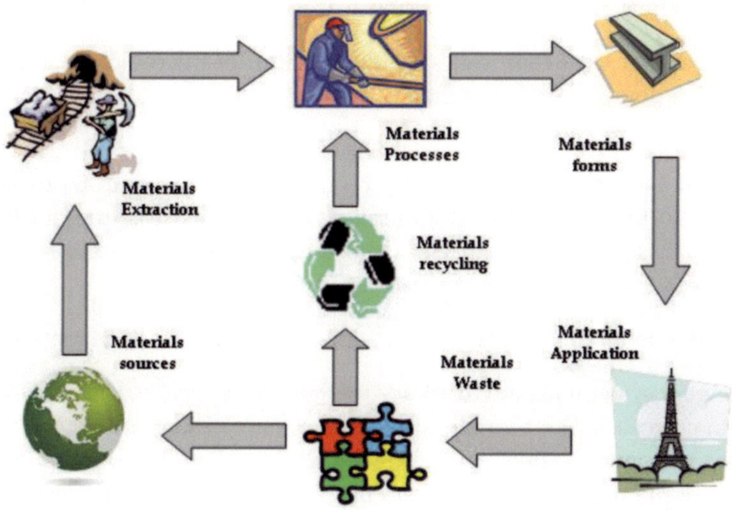

Figure 1. Materials production and use cycle.

In the mineral extraction step, the effluents of N, C, S, NO_x, SO_x and CO_x, from machinery and equipment, operation process water, particulate matter and ground movement in landfills.

The processing stage, chemical operations or extractive metallurgy for converting the concentrate into metal apply selected technologies. The effluents are gases such as SO_2, NO_2 and CO_2, water contaminated with heavy metals, and hazard sediments.

In the manufacturing step the material undergoes operations that transform it into rods, bars, sheets; losses are scrap metal, such as cuts, burrs, mill scale, which recycled with no net loss of metal. In the manufacturing stage the metal is formed by stamping, machining and forging.

Focus on good operations management involves control of air emissions, water management and treatment, solid waste disposal and good land use, will greatly help to maintain a good balance with the environment. It is also necessary to analyze the production area to identify what improvements or measures should be implemented. The role of hydrometallurgist is particularly important and so he is responsible for the design of environmentally friendly processes in each of his steps, to promote sustainable production.

2.1 Processes of materials biodeterioration in industrial systems

In addition to the common processes of deterioration of materials by chemical reactions and mechanical fracture, there are others who are concerned with the participation of various types of microorganisms that adhere in colonies or develop on their surfaces.

Biocorrosion and biodeterioration of metallic materials and nonmetallic materials are two important processes that cause serious problems to the infrastructure of various industrial systems. Generally, microorganisms do not deteriorate or corrode metals directly, but modify the conditions of interface material / environment and surroundings, favoring the degradation of these materials in such a way that induce or influence the development process.

Biofouling is a common term that indicates the presence of microbiological growth on the surfaces of structures built of different materials favoring the formation of biofilms with the colonies of various types of microorganisms.

In the case of metal, biocorrosion occurs due to corrosion electrochemical processes and biological agents due to the action of microorganisms and / or bacteria present in the system. The knowledge of these biological processes and their effects is necessary in order to establish preventive measures and control measures in industrial systems.

An industrial plant containing several biocorrosion environments is a potential risk:

In a heat exchangers system, usually dust accumulates biological waste; biocorrosion could occur, leading to corrosion film formation on walls surface. Therefore, it will be energy loss by increasing the resistance to fluid flow and heat transfer. Loss by evaporation of water favors the increase of the concentration of nutrients, the residence time, the water temperature and the surface / volume ratio, which leads to higher rate of microbial growth (Stoytcheva et al., 2010, Carrillo M. et al., 2010).

Until the early 80's of the twentieth century, we used mixtures of anodic and cathodic inhibitors, such as chromium, zinc and phosphates, to lessen the effects of corrosion in water systems. In some cases we added a polymer, as is still done to date, to avoid or eliminate the problems of fouling on the metal walls. On the other hand, to prevent microbiological growth, we added biocides such as chlorine and quaternary ammonium compounds under acidic conditions.

In the early 90's, the strategies for industrial water treatment changed because of pressure from laws inforcing for the preservation of the environment. Chromates and acid pH values are replaced by the use of organic phosphonates as corrosion inhibitors, while for the control of fouling polycarboxylate type polymers are used. However, this change brought about an increase in the amount of suspended solids, a greater number and variety of microorganisms and therefore a greater amount of inorganic deposits on the heat exchangers walls.

2.2. Biodeterioration of metallic and nonmetallic materials

The metal nature has an effect on the distribution and development of microbial films on its surface. These films influence on the wear and corrosion of the metal substrate. The lack of homogeneity in the biofilm is a precursor of differential aeration processes with formation of differential cell concentration, for example, stainless steels (SS) and nickel-copper (Ni-Cu) alloys in seawater. The oxides passive films or hydrated hydroxides (corrosion products) are a good place for the establishment and growth of bacteria, especially when these products are at a physiological pH values (pH ≈ 7.4)

- Carbon Steel (CS)

CS are very active metals in aggressive media, such as seawater. In this case, the action of microorganisms involves the dissolution of films of corrosion products, by processes of oxidation and reduction. This creates new metal active areas, exposed to the aggressive medium and suffers corrosion processes. In the case of sulfate-reducing bacteria (SRB), the species generated by their metabolism (sulfides) are corrosive to the metal. Figure 2 shows the final state pitting outside a CS pipe, which was affected by microbial growth inside, prompting a process of microbial corrosion with not uniform localized attack.

- Stainless steel

The presence of chromium and molybdenum as alloying elements, enable passive behavior of stainless steels in different environments. However, the passive surface of these SS provides an ideal location for microbial adhesion and therefore are susceptible to corrosion pitting, crevice corrosion under stress or in solutions containing chlorides, as sea water.

Figure 2. External pitting caused by biocorrosion on the internal surface of a carbon steel pipe in a fire extinguisher system.

In marine environments, the generation of peroxides during bacterial metabolism causes an ennoblement of the pitting potential of SS, thus promoting corrosion. Obviously, not all SS have the same behavior, but in general they tend to deteriorated in the presence of colonies of microorganisms.

- Copper and nickel alloys

Alloys of Cu with Zn, Sn and Al, brasses, bronzes, aluminum bronzes; also the nickel alloys: Monel, Hastelloy, nickel superalloys: Ni-Mo, Ni-Cr-Mo, Ni-Cr-Fe- Mo; the traditional nickel alloys: Ni-Cr-Fe, Ni-Fe-Cr, Fe-Ni-Cr-Mo), and the Cu-ni alloys CuNi\70/30, CuNi\90/10, have shown great corrosion resistance in different environments, so they have found a wide use in different industries and environments. However, despite these skills, there are reports that these alloys are colonized by bacteria after several months of exposure in seawater (Acuña, N. et al., 2004).

- Aluminum and its alloys

Al is an active metal which is passivated rapidly in some neutral and acid media, thus offering a good resistance to corrosion. Al alloys with copper, magnesium and zinc, are widely used in the aviation industry. However, there have been cases of biocorrosion on fuel tanks of jet aircraft made of Al alloys by microbial contaminants in turbo combustibles. The presence of water (moisture), even in minimal amounts, allows growth of microorganisms (typically fungi), when these are able to utilize hydrocarbons as a carbon source.

- Titanium

Ti is considered as the most resistant metal to biocorrosion, according to the results of tests carried in different conditions, due to its passive behavior that is reinforced in the presence of oxidizing agents. This is the reason why Ti is the material of choice, for example, for the manufacture of tubes in cooling systems that use seawater.

- Nonmetallic materials

Non-metallic materials such as fiberglass reinforced polyester (FGRP), concrete and wood, are also affected by biodeterioration processes in the presence of microorganisms

In the case of FGRP, bacteria and algae are able to use the polyester matrix as a carbon source, consuming and considerably reducing the mechanical strength of composite material, ultimately causing its failure. This is easily observable in screens of this material in cooling towers or tanks containing fresh water or salt water. Wood suffers biodeterioration by the presence of fungi in moist environments that promote the delignification of this material (Valdez B., et al., 1996, 1999, 2008).

2.3. Facing the problems of biodeterioration

The inevitable presence of microorganisms in the feed water causes a sequence of biofouling, biocorrosion and biodeterioration of the materials component of the structures. This sequence depends on the degree of microbial contamination and the system operating characteristics.

The most common methods of controlling these problems involve the application of continuous or metered biocides such as chlorine. Currently, we use substances more compatible with the environment, since the use of chlorine is limited to certain concentrations. Such is the case of ozone, which is also ascribed with passivating effects on certain metals and alloys commonly applied in industry, and also in antifouling action.

In order to tackle a biodeterioration problem it is required a prior analysis of the problem, to know when conditions are suitable for the development of this process. In industrial systems we need to know some parameters: temperature, pH, nutrients; carbon, phosphorus, nitrogen, sulfate ion levels and flow rates. The places where we find biodeterioration are: biofouling deposits, under any deposit, zones of localized metal corrosion. to check their presence it is necessary to utilize sampling techniques, isolation and identification of microorganisms. It is interesting to note that there are commercial devices for in situ measurements that are practical and useful for the plant engineer.

3. Corrosion in the electronics industry

Corrosion of device components, manufactured by the electronics industry, is a problem that has occurred during a long time. Often, especially corrosion of one or more of the metallic elements of an electronic component is the primary cause of failure in various electronic equipments. The high density of components required to reduce the size of electronic equipment, also for a better signal processing, leads to the generation of enclosed corrosion between thin metal sections. Furthermore, when electronic devices are in more severe environments such as tropical, subtropical, contaminated deserts, etc., they have high failure rates. Problems, due to the aggressiveness of the medium in electronic equipment for military use, have also occurred in aircraft and submarine guidance systems. Another common problem is corrosion damage suffered by components music players, when exposed to humid environments contaminated with chlorides, for example, during transport by ship, from the manufacture location to the consumer place. Thin layers of corrosion products on the surface of the metal component change their electrical characteristics: resistance, capacity and lead to partial or total failure of the electronic system. There are reported cases where small amounts of moisture have caused corrosion in tablets with printed circuits, nichrome resistors, fittings, electrical connectors and a wide range of components, and micro-electronic components, which have been coated with metallic films (Valdez B. et al., 2006, G. Lopez et. al., 2007)

Corrosion of metal components in the electronics industry may occur at different stages: during manufacture, storage, shipping and service. The main factors in the onset of corrosion and subsequent development are moisture and corrosive pollutants, such as chlorides, fluorides, sulfides and nitrogen compounds, organic solvent vapors, emanating from the resins used as label, or coatings formed during the curing process and packaging of microcircuits.

The sources providing aggressive pollutants are diverse, from flux residues used for welding processes, waste and vapors from electrolytic baths, arising volatile organic adhesives, plastics and acidification of their environment. Assays in artificial atmosphere, which simulates an indoor environment of an electronic plant have shown that the surface of the silver undergoes browning or tarnishing and the formation of dendrite whiskers due to corrosion (Figure 3).

The elemental chemical analysis of the surface (EDX - Scattered Electron Spectroscopy and XRD - X-rays) shows that the corrosion product formed on the silver surface is silver sulfide (Ag_2S), due to the action of pollutant gases such as SO_2 and H_2S present in a humid environment (Figure 4). Moreover, the micrograph of the silver surface (SEM) shows a dendritic growth of corrosion products, characteristic for silver components.

The design of electronics equipment requires a great variety of different metals, due to their different physical and electrical features. Metals and alloys used in the electronics industry are:

- Gold (Au) coating and / or foil in electrical connectors, printed circuits, hybrid and miniature circuits.;
- Silver (Ag) for protective coating in contact relays, cables, EMI gaskets, etc..;
- Magnesium (Mg) alloys for radar antenna dishes and light structures, chassis brackets, etc..;
- Iron (Fe), steel and ferroalloys for guide components, magnetic shielding, magnetic coatings memory disks, processors, certain structures, etc..;
- Aluminum (Al) alloys for armor equipment, chassis, mounting frames, brackets, trusses, etc..;
- Copper and its alloys for cables, tablets printed circuit terminals, nuts and bolts, RF packaging, etc..;
- Cadmium (Cd) for sacrificial protective coating on iron and safe electrical connectors;
- Nickel (Ni) coating for layers such as barrier between copper and gold electrical contacts, corrosion protection, electromagnetic interference applications and compatibility of dissimilar material joints;
- Tin (Sn) coating for corrosion protection of welding; for compatibility between dissimilar metals, electrical connectors, RF shielding, filters, automatic switching mechanisms;
- Welding and weld coatings for binding, weldability, and corrosion protection.

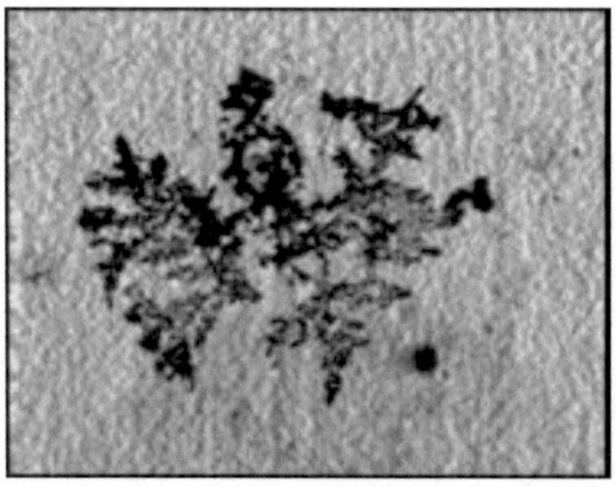

Figure 3. Silver sulfide whiskers corrosion products on silver exposed in an electronics plant atmosphere.

Many of these metals are in contact with each other, so that in the presence of moisture, galvanic corrosion / bimetallic corrosion occurs. When using similar metals, due to design the following requirements must be taken into account.

- Designing the contact of different metals such that the area of the more noble cathodic metal should be appreciably smaller than the area of the more active anodic metal. The area of the cathode can be decreased by applying paint or coating.
- Coating the contact area of a metal with a compatible metal.
- Interpose between dissimilar metals in a metal compatible packaging.
- Sealing interfaces to prevent ingress of moisture.
- Set the electronic device in a hermetically sealed arrangement.

Other corrosion problems can occur due to the characteristics of electronic components such as electromagnetic interference, electromagnetic pulse, flux residues, finishes and materials component tips, organic products that are used for various purposes and emitting gases during curing, whiskers, embrittlement inter-metallic electrical contacts.

Metal components may corrode during manufacture and storage prior to assembly, needing protection against corrosion. In plants and warehouses, air conditioning systems must operate efficiently, removing moisture and suspended particulate matter. Filters and traps should be cleaned and replaced regularly. For closed containers, we recommend the installation of dryers with visual indicators, and the use of volatile vapor phase corrosion inhibitors. In the case of sealed black boxes, the temperature inside these drops should never be below the dew point (Veleva L. et al., 2008, Vargas L. et al., 2009, Lopez G. et al., 2010).

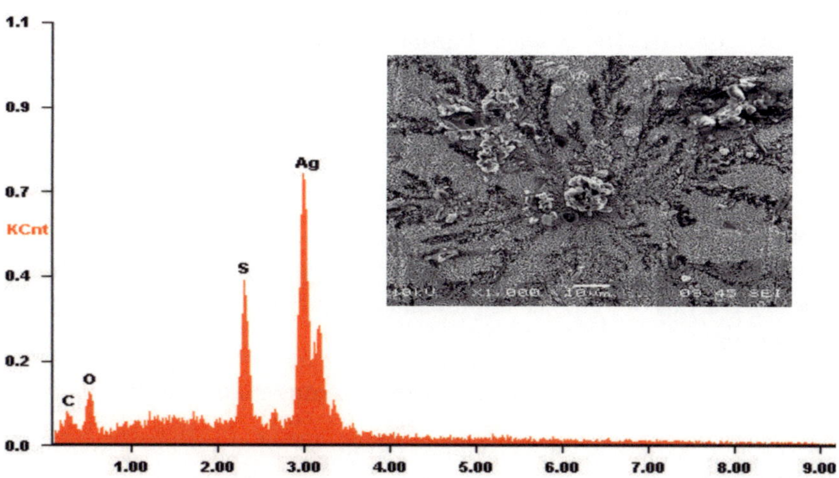

Figure 4. Scanning Electron Microscopy and EDS analysis of silver corrosion products at indoor conditions of an electronics plant contaminated with H_2S.

4. Corrosion in water

Abundant water sources are essential to a country's industrial development. Large quantities of this precious liquid are required for cooling products, machinery and equipment, to feed boilers, meet health needs and provide drinking water to humans. Estimates of water consumption for each country are different and depend on the degree of industrial development thereof. In first world countries like the United States, these intakes are as high as several hundred billion liters per day. These countries have implemented water reuse systems with certain efficiency due to the application of appropriate treatment for purification. Water, a natural electrolyte is an aggressive environment for many metals / alloys, so that they may suffer from corrosion, whose nature is electrochemical.

As raw water or fresh water we mean natural water from direct sources such as rivers, lakes, wells or springs. Water has several unique properties and one of these is its ability to dissolve to some degree the substances found in the earth's crust and atmosphere allowing the water to contain a certain amount of impurities, which causes problems of scale deposition on the metal surface, e.g. in pipelines, boiler tubes and all kinds of surfaces that are in contact with water (Valdez, B. et al., 1999, 2010).

Oxygen is the main gas dissolved in water, it is also responsible for the costly replacement of piping and equipment due to its corrosive attack on metals in contact with dissolved oxygen (DO). The origin of all sources of water is the moisture that has evaporated from the land masses and oceans, then precipitated from the atmosphere. Depending on weather conditions, water may fall as rain, snow, dew, or hail. Falling water comes into contact with gases and particulate matter in the form of dust, smoke and industrial fumes and volcanic emissions present in the atmosphere.

The concentrations of several substances in water in dissolved, colloidal or suspended form are low but vary considerably. A water hardness value greater than 400 parts per million (ppm) of calcium carbonate, for example, is sometimes tolerated in the public supply, but 1 ppm of dissolved iron should be unacceptable. In treated water for high pressure boiler or where radiation effects are important, as in nuclear reactors, impurities are measured in very small amounts such as parts per billion (ppb).

In the case of drinking water the main concern are detailed physicochemical analysis, to find contamination, and biological assays to detect bacterial load. For industrial water supplies it is of interest the analysis of minerals in particular salts. The main constituents of water are classified as follows:

- Dissolved gases: oxygen, nitrogen, carbon dioxide, ammonia and sulfide gases;
- Minerals: calcium, sodium (chloride, sulfate, nitrate, bicarbonate, etc.), Salts of heavy metals and silica;
- Organic matter: plant and animal matter, oil, agricultural waste, household and synthetic detergents;

http://dx.doi.org/10.5772/51987

• Microbiological organisms: include various types of algae, slime forming bacteria and fungi.

The pH of natural waters typically lies within the range of 4.5 to 8.5; at higher pH values, there is the possibility that the corrosion of steel can be suppressed by the metal passivation. For example, Cu is greatly affected by the pH value in acidic water and undergoes a slight corrosion in water releasing small amounts of Cu in the form of ions, so that it's corroded surface because green stained clothing and sanitary ware. Moreover, deposition of the Cu ions on surfaces of aluminum or galvanized zinc corrosion cells leads to new bimetallic contact, which cause severe corrosion in metals.

The mineral water saturation produces a greater possibility of fouling on the metal walls, due to the ease with which the insoluble salts (carbonates) can be precipitated. To control this effect it is necessary to know and use the Saturation Indices. Water saturation refers to the solubility product of a compound and is defined as the ratio of the ion activity and the solubility product. For example, water is saturated with calcium carbonate when it is no more possible to dissolve the salt in water and then it begins to precipitate as scale. In fact, it is called supersaturated when carbonate precipitation occurs on standing the solution. The most common parameters that must be known to characterize the water corrosivity, be it raw or treated, for operation in an industrial facility are shown in Table 2.

Water properties	Corrosivity
Hardness	Source of scaling that promotes corrosion
Alkalinity	Produces foam and motion of solids
pH	Corrosion depends on its value
Sulphates	Produces scaling
Chloride	Increases water corrosivity
Silica	Generates scaling in hot water. Condensers and steam turbines
Total Dissolved Solids (TDS)	Increases electrical conductivity and corrosivity
Temperature	Elevated temperatures increases corrosion rates

Table 2. Water properties and corrosivity.

There six formulas to calculate Saturation Indices and embedding: Langelier index (LSI), Ryznar stability index, Puckorious index of scaling, Larson-Shold index, index of Stiff- Davis and Oddo-Tomson index. There is some controversy and concern for the correlation of these indices with the corrosivity of the waters, particularly regarding the Langelier (LSI).

A LSI saturation index with value "0" indicates that the water is balanced and will not be fouling, while the positive value indicates that the water may be fouling (Table 3). The negative value of the LSI suggests that water is corrosive and can damage the metal installation, increasing the content of metallic ions in water. While some sectors of the water management industry uses the values of the indices as a measure of the corrosivity of the water. Corrosion specialists are alerted and are very wary of issuing an opinion, or extrapolate the use of indices to measure the corrosivity of the environment.

Langelier Index	Water corrosivity and scaling
-5.0	Severe corrosion
-4.0 to -2.0	Moderate corrosion
-1.0 to -0.5	Light corrosion
0.0	No corrosion / no scale (balance)
0.5 to 2.0	Light incrustation
3.0	Moderate incrustation
4.0 to 5.0	Severe incrustation

Table 3. Langelier index for water corrosivity and scaling.

Sometimes the raw water is contaminated with chemicals such as fertilizers and other chemicals coming from agricultural areas (Figure 5).

In these cases, ionic agents such as nitrites, nitrates, etc., in water causes an accelerated process of localized corrosion to many metals and the consequent failure of equipment.

Figure 5. Corrosion on the gates dam on the Colorado River, Baja California, Mexico

Raw water contaminants can be quite varied, including both heavy metals and organic chemicals, referred to as toxic pollutants. Among the heavy metals may be mentioned arsenic (As), mercury (Hg), cadmium (Cd), lead (Pb), zinc (Zn) and cadmium (Cd), which are sometimes at trace levels, but they tend to accumulate over time, so that priority pollutants are to be treated.

Pesticides, insecticides and plaguicides comprise a long list of compounds, for which we should be concerned: DDT (insecticide), aldrin (an insecticide), chlordane (pesticide), endosulfan (insecticide), diazinon (insecticide), among others.

Contaminants, such as polycyclic aromatic organic compounds, include what is known as volatile organic compounds such as naphthalene, anthracene and benzopyrene. There are two main sources of these pollutants: petroleum and combustion products found in municipal effluents. On the other hand, there are polychlorinated biphenyls or PCBs, which are mainly used in transformers for the electrical industry, heavy machinery and hydraulic equipment. This class of chemicals is extremely persistent in the environment and affects human health.

From the viewpoint of corrosion, these contaminants which are present even at low concentrations or trace in the raw water, favor the corrosivity the metals which are in contact with. The combination of the corrosive effects of these contaminants together with the oxidation by oxygen, minerals and other impurities, leads to consider raw water as a natural means capable of generating corrosion of metals. It is recommended at least, to carry out a process of treating raw water, to reduce significantly the hardness and remove suspended solids, which will help greatly in preventing subsequent problems of corrosion and fouling on metal surfaces, curbing economic losses and maintaining the industrial process in good operating condition.

4.1. Corrosion in potable water systems

Corrosion is a complex phenomenon that arises as a result of the interaction between water and the surface of metallic pipes or the equipment of storage and handling. The process is invariably a combination of oxidation and reduction, as already described in previous chapters. In drinking water, it should be noted that the corrosion products which are partially soluble in water in ionic form are toxic at certain concentrations, e.g. copper and lead. The existence of high concentrations of lead in water carried by copper tubing, indicate that the source of lead may be tin-lead solder at the junctions of the copper pipes. The consumption of domestic water contaminated with toxic metal ions (Pb^{+2}, Cu^{+2}, Zn^{+2}, Cr^{+3}), gives rise to acute chronic health problems. The regulations have set the following limits allowable concentration in drinking water: Cr (0.05 ppm), Cu (0.01 ppm), Pb (0.05 ppm) and Zn (5 ppm). These regulations are made in order to protect the public user and consumer of drinking water and are continuously striving for a reduction in the maximum allowable limits. Some concentrations reach zero as is the case of Pb in the United States due to the concerns Pb about poisoning of children. Still, many sources such as wells and springs are outside the control of law and toxic substances, bacteria and pathogens. Damage caused by corrosion of household plumbing may be accompanied by unpleasant aesthetic problems such as soiled clothing, unpleasant taste, stains and deposits in the toilets, floors of bathrooms, tubs and showers. To prevent corrosion of pipes, we recommend the use of PVC pipes for drinking water, replacing the metal, as a preventive measure.

Corrosion can occur anywhere on the pipes that carry drinking water, mainly at sites of contact between two dissimilar metals, thus forming a corrosion cell. In general, the metals will corrode to a greater or lesser degree in water, depending on the nature of the metal, on the ionic composition of water and its pH. Waters high in dissolved salts (water hardness), favor the formation of scale, more or less adherent, in different parts of the equipment (Figure 6). These deposits may be hard or brittle, sometimes acting as cement, creating a physical barrier between the metal and water, thereby inhibiting corrosion. Calcium carbonate ($CaCO_3$) is the most common scale; its origin is associated with the presence of carbon dioxide gas (CO_2) in water. Sometimes these deposits are filled with pasty or gelatinous hydrated iron oxides or colonies of bacteria (Valdez, B. et al., 1999, 2010).

Figure 6. Corrosion in potable water pipes.

Usually, groundwater $CaCO_3$ saturated (calcareous soils), due to the presence of dissolved CO_2, whose content depends on its content in the air in contact with the water and on temperature. These waters are often much higher in CO_2 content, so they may dissolve substantial amounts of calcium carbonate. These waters are at pressures lower than they had in the ground, so CO_2 gas lost with consequent supersaturation of carbonates. If conditions are appropriate, the excess of $CaCO_3$ can precipitate as small agglomerates deposited in muddy or hard layers on solid surfaces, forming deposits. An increase in temperature is an important factor and also leads to supersaturation of carbonates, with the consequent possibility of fouling. To a lesser extent fouling can precipitate more soluble Mg carbonates ($MgCO_3$) and Mn ($MnCO_3$), and also oxides / hydroxides, dark colored and gelatinous. Except in very exceptional cases in sulfated water, it is normal to find deposits of gypsum ($CaSO_4 \bullet ½ H_2O$) because their solubility is high, but decreases with increasing temperature. Hard silica scale (SiO_2) may appear with oversaturated waters or appear as different silicates (SiO_4^{4-}) trapped in the carbonate deposits. Generally, the silica appears trapped in other types of scale and it is not chemical precipitation.

Waters often carry considerable amounts of iron (ferrous ion, Fe^{+2}), which may be often precipitated by oxidation upon contact with air as hydrated iron oxide (ferric, Fe^{+3}) but sometimes can be Fe^{+2} form black sludge, more or less pasty or gelatinous and sometimes very large. The voluminous precipitate occupies the pores, significantly reducing the permeability of the fouling. Sometimes the Fe ions can come from corrosion of the pipe giving rise to simultaneous corrosion and scaling (Figure 6). Common bacteria of the genera Gallionella, Leptothrix Cremothrix are known as Fe bacteria, can give reddish-yellow voluminous precipitate and sticky ferric compounds from ferrous ion, which drastically reduce the permeability of the deposit, in addition to trap other insoluble particles.

The cost for impairment of domestic water systems and the impact on health, involves several consequences: premature corrosion and failure of the pipes and fittings that carry water in a house or building, a low thermal efficiency (up to 70%) of water heaters (boilers), which can cause their premature failure. High levels of metals or oxides, which usually are not properly, treated in drinking water cause red or blue-green deposits and stains in the toilets sinks. In addition to concerns about the aesthetic appearance, a corrosion process can result in the presence of toxic metals in our drinking water. For evaluating water quality and their tendency corrosive and / or fouling, LSI can be used. This analysis must be accompanied by measurements of water pH and conductivity, and corrosion tests applying international standards.

4.2. Anticorrosive treatment of water

Corrosion control is complex and requires a basic knowledge of corrosion of the system and water chemistry. Systems can be installed for water pretreatment, using non-conductive connections, reducing the temperature of hot Cu water pipes employed and copper installing PVC or other plastic materials. It is important to note that the corrosiveness of water can be increased by the use of water softeners, aeration mechanisms, increasing the temperature of hot water, water chlorination, and attachment of various metals in the water conduction system. A proper balance between the treatment systems and water quality, can be obtained with acceptable levels of corrosivity. Thus, the lifetime of the materials that make the water system in buildings, public networks, homes and other systems will be longer.

5. Soil corrosion

A large part of steel structures: aqueducts, pipelines, oil pipelines, communications wire ropes, fuel storage tanks, water pipes, containers of toxic waste, are buried, in aggressive soils. Large amounts of steel reinforced concrete structures are also buried in various soil types. In the presence of soil moisture it is possible to have humid layer on the metal surface, whose aggressiveness depends on soil type and degree of pollution (decaying organic matter, bacterial flora, etc.). Thus, the soil can form on the metal surface an electrolyte complex with varying degrees of aggressiveness, a necessary element for the development of an underground electrochemical corrosion. The corrosion process of buried structures is extremely variable and can occur in a very fast, but insignificant rate, so that pipes in the soil can have perforations, presenting localized corrosion attack or uniform.

Metal structures are buried depending on their functionality and security. Most often they traverse large tracts of land, being exposed to soils with different degrees of aggressiveness exposed to air under atmospheric conditions (Figure 7).

Figure 7. Valve system of a desert water aqueduct.

When pipes or tanks are damaged by corrosion, the formation of macro-and micro-cracks can lead to leaks of contained products or fluids transported, causing problems of environmental pollution, accidents and explosions, which can end in loss of life and property (Guadalajara, Jalisco, Mexico, 1992). In the case of pipes used to carry and distribute water, a leak may cause loss of this vital liquid, so necessary for the development of society in general and especially important in regions where water is scarce, so the leakage through aqueducts pipes should be avoided. An important tool needed to prevent the most serious events, is the knowledge of the specific soil and its influence on the corrosion of metal structures.

5.1. Types of soils and their mineralogy

A natural soil contains various components, such as sand, clay, silt, peat and also organic matter and organisms, gas, mineral particles and moisture. The soils are usually named and classified according to the predominant size range of individual inorganic constituent particles. For example, sandy soil particles (0.02 - 2 mm) are classified as fine sand (0.02 - 0.2 mm) or thick (0.20 -2.00 mm). Silt particles (0.002 to 0.02 mm) and clay, which have an average diameter 0.002 mm, are classified as colloidal matter. A comparison of the sizes of these typical soils is done in Figure 8.

Currently exists in the U.S. and in over 50 countries worldwide, a detailed classification for soils, which includes nine classes with 47 subgroups.

The variation in the proportion of the groups of soil with different sizes, determines many of its properties. Fine-textured soils due to high clay content, have amassed particles, so they have less ability to store and transport gases such as oxygen, that any ground-open e.g. san-

http://dx.doi.org/10.5772/51987

dy soil. The mineralogy of both clay types and their properties, are closely related to the corrosivity of the soil. Silica (SiO_2) is the main chemical constituent of soils type clay, loam and silt, also in the presence of Al_2O_3. Common species in moist soil are dissolved ions H^+, Cl^-, SO_4^{2-}, HCO_3^-. The chemical composition and mineralogy of the soil determine its corrosive aggressiveness; poorly drained soils (clay, silt and loam) are the most corrosive, while soils with good drainage (gravel and sand type) are less aggressive to metals. Vertically homogeneous soils do not exist, so it is convenient to consider the non-uniformity of ground, formed of different earth layers. To understand the corrosion behavior of a buried metal is very important to have information about the soil profile (cross section of soil layers). The physicochemical and biological nature of soil, corrosive aggressiveness and dynamic interactions with the environment, distinguishes the ground like a very complex environment and different from many others. Climate changes of solar radiation, air temperature and relative humidity, amount of rainfall and soil moisture are important factors in corrosion. Wind, mechanical action of natural forces, chemical and biological factors, human manipulation can alter soil properties, which directly affects the rate of corrosion of metals buried in the ground. Conditions may vary from atmospheric corrosion, complete immersion of the metal, depending on the degree of compactness of the soil (existence of capillaries and pores) and moisture content. Thus the variation in soil composition and structure can create different corrosion environments, resulting in different behavior of the metal and oxygen concentrations at the metal / soil interface.

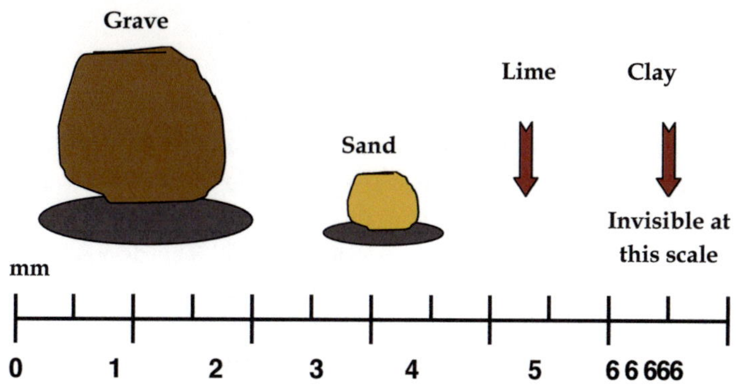

Figure 8. Size of soil particles.

Two conditions are necessary to initiate corrosion of metal in soil: water (moisture, ionic conductor) and oxygen content. After startup, a variety of variables can affect the corrosion process, mentioned above, and among them of importance are the relative acidity or alkalinity of the soil (pH), also the content and type of dissolved salts.

Mainly three types of water provide moisture to the soil: groundwater (from several meters to hundreds below the surface), gravitational (rain, snow, flood and irrigation) and capillary (detained in the pores and capillary spaces in the soil particles type clay and silt). The moisture content in soils can be determined according to the methodology of ASTM D 2216

("Method for Laboratory Determination of Water (Moisture) Content of Soil and Rock by Mass"), while its permeability and moisture retention can be measured the methods described in ASTM D2434 and D2980. The presence of moisture in soils with a good conductivity (presence of dissolved salts), is an indication for high ion content and possible strong corrosive attack.

The main factors that determine the corrosive aggressiveness of the soil are moisture, relative acidity (pH), ionic composition, electrical resistance, microbiological activity.

5.2. Corrosion control of buried

Given the electrochemical nature of corrosion of buried metals and specific soils, this can be controlled through the application of electrochemical techniques of control, such as cathodic protection. This method has been universally adopted and is appropriate to protect buried metallic structures. For an effective system of protection and cheaper maintenance, pipelines must be pre-coated, using different types of coatings, such as coal tar, epoxies, etc. This helps reduce the area of bare metal in direct contact with the ground, lowering the demand for protection during the corrosion process. The purpose of indirect inspection is to identify the locations of faulty coatings, cathodic protection and electrical Insufficient shorts (close-interval, on/off Potential surveys, electromagnetic surveys of attenuation current, alternating current voltage gradient surveys, etc..), interference current, geological surveys, and other anomalies along the pipeline.

6. Corrosion under thermal insulation

One of the most common corrosion problems in pipes, ducts, tanks, preheaters, boilers and other metal structures, insulated heat exchange systems, is the wear and corrosion occurring on metal (steel, galvanized steel, Al, SS, etc.), below a deposit or in its immediate neighborhood. This corrosion is known as corrosion under deposit. The deposit may be formed by metal corrosion products and / or different types of coating applied for protection. For example, in the case of a calcareous deposit, formed in the walls of galvanized steel pipes which carry water with a high degree of hardness (dissolved salts), it might develop corrosion under deposit. These shells may be porous, calcareous deposit and / or partially detached from the metal surface, so that direct contact between metal, water and oxygen (the oxidizing agent in the corrosion process) allows the development of metal corrosion. For this reason the pipes could be damaged severely in these locations up to perforation, while in parts of the installation corrosion might occur at a much lower level.

There is a considerable amount of factors in the design, construction and maintenance, which can be controlled to avoid the effects of deterioration of metal by corrosion under deposit. In general, under these conditions the metal is exposed to frequent cycles of moisture, corrosivity of the aqueous medium or failure in the protective coatings (paint, metal, cement, fiberglass, etc.). Figure 9 shows a conductor tube steam in a geothermal power plant, where CS corrosion happened beneath the insulation.

http://dx.doi.org/10.5772/51987

Figure 9. Corrosion of a carbon steel pipe under insulation.

Seven factors can be controlled on the ground, to prevent this type of corrosion: design of equipment, operating temperature, selection of the insulation, protective coatings and paints, physical barriers from the elements, climate and maintenance practices of the facility. Any change in any of these factors may provide the necessary conditions for the corrosion process to take place. The management knowledge of these factors help explain the causes of the onset conditions of corrosion under deposits, and it will guide a better inspection of existing equipment and the best design.

6.1. Equipment design

The design of pressure vessels, tanks and pipes, generally includes accessories for support, reinforcement and connection to other equipment. Details about the installation of accessories are the responsibility of the engineers or designers, using building codes to ensure reliability of both insulated and non insulated equipment. The protective barrier against the environment surrounding the metal structure in such designs often breaks donor due to an inappropriate insulation, loss of space for the specified thickness of insulation or simply by improper handling during installation of the equipment. The consequence of a rupture or insulation failure means greater flow water ingress to the space between metal and coating hot-cold cycle, generating over time a buildup of corrosive fluid, increasing the likelihood of corrosive damage. Moreover, wet insulation will be inefficient and also cause economic losses. The solution of this factor is to meet the thickness specifications and spacing, as indicated in the code or equipment-building specifications and characteristics of the coating used.

The operating temperature is important for two reasons: a high temperature favors the water is in contact with the metal for less time, however, also provides a more corrosive environment, causes fast failures of coatings. Usually a team operating in freezing temperatures is protected against corrosion for a considerable life time. However, some peripheral devices, which are coupled to these cold spots and operating at higher temperatures, are exposed to moist, air and steam, with cycles of condensation in localized areas, which make them more vulnerable to corrosion. For most operating equipment at freezing conditions, the corrosion occurs in areas outside and below the insulation. The temperature range where this type of corrosion occurs is 60 °C to 80 °C; however, there have been failures in zones at temperatures up to 370 °C. Also, in good water-proof insulation, corrosion is likely to occur at points where small cracks or flaws are present, so that water can reach the hot metal and evaporate quickly. On the other hand, in machines where the temperature reaches extreme values, as in the case of distillation towers, it is very likely to occur severe corrosion problems.

6.2. Selection of insulation

The characteristics of the insulation, which have a greater influence on the corrosion processes deposits, are the ability to absorb water and chemical contribution to the aqueous phase. The polyurethane foam insulation is one of the most widely used; however, in cold conditions they promote corrosion due to water absorption present. The coatings of glass fiber or asbestos can be used in these conditions, always when the capacity of absorbing water do not becomes too high. Corrosion is possible under all these types of coating, such insulation. The selection of insulation requires considering a large group of advantages and disadvantages regarding the installation, operation, cost, and corrosion protection, which is not an easy task. The outside of the insulation is the first protective barrier against the elements and this makes it a critical factor, plus it is the only part of the system that can be readily inspected and repaired by a relatively inexpensive process. The durability and appearance, melting point fire protection, flame resistance and installation costs are other important factors that must be taken into account together with the permeability of the insulation. Usually the maintenance program should include repairs to the range of 2 to 5 years. Obviously the weather is important and corrosion under thermal insulation will more easily in areas where humidity is high. Sometimes conditions of microclimate can be achieved through the use of a good design team.

7. Corrosion in the automotive industry

One of the most important elements of our daily life, which has great impact on economic activity, is represented by automotive vehicles. These vehicles are used to transport people, animals, grains, food, machinery, medicines, supplies, materials, etc. They range from compact cars to light trucks, heavy duty, large capacity and size. All operate mostly through the operation of internal combustion engines, which exploit the heat energy generated by this process and convert it in a mechanical force and provide traction to these vehicles.

The amount and type of materials used in the construction of automotive vehicles are diverse, as the component parts. They are usually constructed of carbon steel, fiberglass, aluminum, magnesium, copper, cast iron, glass, various polymers and metal alloys. Also, for aesthetic and protection against corrosion due to environmental factors, most of the body is covered with paint systems, but different metal parts are protected with metallic or inorganic coatings.

Corrosion in a car is a phenomenon with which we are in some way familiar and is perhaps for this reason that we often take precautions to avoid this deterioration problem.

A small family car, with an average weight of 1000 kg, is constructed of about 360 kg of sheet steel, forged steel 250 kg, 140 kg cast iron mainly for the engine block (now many are made of aluminum), 15 kg of copper wires, 35 kg and of plastic 50 kg of glass that usually do not deteriorate, and 60 kg for rubber tires; which wear and tear. The remaining material is for carpets, water and oil. Obviously, that is an advanced technology in the car industry, with automobiles incorporating many non-metallic materials into their structure. However, the problem of corrosion occurs at parts where the operation of the vehicle is compromised. Corrosion happens in many parts of the car (mostly invisible) it is not only undesirable for the problems it causes, but also reduces the vehicle's resale value and decreases the strength of the structure. To keep the car in good condition and appearance, its high price, it is necessary to pay attention to the hidden parts of the vehicle.

The main cause of corrosion of the car body is the accumulation of dust in different closed parts, which stays for a long time by absorbing moisture, so that in these areas metal corrosion proceeds, while in the clean and dry external parts it does not occur (Figure 10).

The corrosion problem that occurs in the metal car body has been a serious problem that usually arises most often in coastal environments, contaminated with chlorides and rural areas with high humidity and specific contaminants. Many countries use salt (NaCl, $CaCl_2$ or $MgCl_2$) to keep the roads free of ice; under these conditions these salts, in combination with the dust blown by the car, provide conditions for accelerated corrosion. Therefore, it is recommended as a preventative measure, after a visit on the coast or being on dirty roads, to wash the car with water, and also the tires and the doors, especially their lower parts. In urban environments, the corrosion problem has been reduced due to the new design and application of protective coatings, introduced by major manufacturers in the early nineties of the twentieth century. The areas most affected are fenders, metal and chrome bumpers views which are used in some luxury vehicles as well as areas where water and mud are easily accumulated e.g. auctions of funds windshield and doors (Figure 11).

In regions with high incidence of solar radiation and the presence of abrasive dust, paint vehicles deteriorate rapidly. The hot, humid weather, combined with high levels of SO_2 and NO_x emissions that come from burning oil, chlorides salt. In the Gulf of Arabia, the blowing sand from the nearby desert, creates a very aggressive environment; statistics reveals that one in seven cars is damaged and due to corrosion the car life is estimated to an average of 8 months, also the car corrosion resistance decreases in the following order: manufactured in Europe, USA and Japan. White paints generally have shown a significantly better corrosion protection than other colors. Initially, corrosion defects appear as a kind of dots and spots of corrosion products formed under the paint and subsequently emerge from the steel sheet, leaving a free entry for moisture and air (oxygen), accelerating the corrosion process; in these cases reddish metal corrosion products.

Figure 10. Corrosion on a bodywork exposed to the Gulf of Mexico tropical coast.

Figure 11. Corrosion on a car door and bottom of the bodywork

http://dx.doi.org/10.5772/51987

7.1. Corrosion in the cooling system

The cooling system of a car combustion engine consists of several components, constructed of a variety of metals: radiators are made of copper or aluminum, bronze and solder couplings with tin water pumps; motors are made of steel, cast iron or aluminum. Most modern automobiles, with iron block engine and aluminum cylinder head, require inhibitor introduced into the cooling water to prevent corrosion in the cooling system. The inhibitor is not antifreeze, although there are in the market solutions which have the combination of inhibitor-antifreeze. The important thing is to use only the inhibitor recommended in the automobile manual and not a mixture of inhibitors, since these may act in different ways and mechanisms. The circulating water flow should work fine without loss outside the system. If the system is dirty, the water should be drain and filling the system with a cleaning solution. It is not recommended to fill the system with hard water, but with soft water, introducing again the inhibitor in the correct concentration. If there exhaust at the water cooling system, every time water is added the inhibitor concentration should be maintained to prevent.

In small cars, it is common for water pumps; constructed mainly of aluminum, to fail due to corrosion, cavitation, erosion and corrosion, making it necessary to replace the pump (Valdez, B. et al., 1995). Accelerated corrosion in these cases is often due to the use of a strong alkaline solution of antifreeze. On the other hand, in heavy duty diesel trucks, the cooling system is filled with tap water or use filters with rich conditioner chromates that can cause the pistons jackets to suffer localized corrosion. After 12 or 15 months, the steel jackets are perforated and the water passes into the cavity through which the piston runs, forcing to carry out repair operations (Figure 12).

Figure 12. Corrosion in a carbon steel jacket on the water face in a diesel combustion engine truck.

Corrosion causes great economic losses to the transport industry, since it must stop to repair the truck and abandon to provide the service with all the consequences that this entails. Furthermore, the use of chemical conditioning is now controlled by environmental regulations, so chromates and phosphates are restricted and novel mixtures of corrosion inhibitors have been produced to control the problem of corrosion in automobile cooling systems.

7.2. Corrosion in exhaust pipes and batteries

Exhaust pipes made of SS (0.6 - 0.8 mm thick) have a better resistance to chemical corrosion at high temperatures, which is why we are now using SS in many popular models. This SS resists corrosion much more than conventional CS and thus their long life covers the higher price. Another alternative is to use conventional CS tube, zinc coated or aluminum (Figure 13). These exhaust pipes are less expensive than stainless steel, but less resistant to corrosion.

Figure 13. Corrosion on carbon steel exhaust pipes coated with aluminum.

The acidic environment which is generated on the surface of accumulators supplying the energy necessary for starting the engine, favors conducting corrosion processes in the lead terminals, where the cables are connected by bronze or steel clamps. Thus, this environment and these contact zones predispose cells to a process galvanic corrosion, which gradually deteriorates the contact wires, generating bulky corrosion products. This phenomenon is called sulfation of the contacts due to the sulfuric acid containing the battery, thus forming white sulfates on the corroded metal surface. These products introduce high resistance to current flow and cause failure to the engine ignition system, and impede the battery charge

http://dx.doi.org/10.5772/51987

process. This problem has been eliminated in batteries that have airtight seals, or are manufactured with new technologies as well as bases covered with organic coatings that prevent corrosion.

Some years ago it was common for starters to fail, because the moisture or water penetrated into the gear area preventing it sliding motion and causing burning of the electric motor. Currently, new designs avoid contact with moisture and other foreign agents, preventing the occurrence of corrosion problems in these devices. As a preventive measure is recommended to prevent spillage of battery acid, to periodically clean the battery terminals (with a brush of wire or a special instrument), also coat them with petroleum jelly to prevent corrosion in these contact areas. A fat based composition which contains several components: alkaline salts and oxides of lithium, sodium bicarbonate and magnesium oxide are applied to the terminals and the connector. In general, in wet weather, the contacts of the accumulators have a tendency to more accelerated corrosion, thus requiring greater care to disconnect the terminals when not being used.

7.3. Corrosion prevention

To keep the vehicle for a longer time without the appearance of corrosion, it always requires washing with running water and, the use of very soft brush or cloth-like material, with a special detergent (not household detergents, which are very corrosive) and finally wash the vehicle with plenty of water. The floor carpet should be maintained clean and dry. A car should not be left wet in a hot garage, since under these conditions accelerated corrosion takes place since the water does not dry and can condense on the cold parts of the vehicle. In these cases, it is best not to close the garage door or use a roof space, to protect it from rain, and not allow moisture condensation. However, if the vehicle is left unused for a long time in a closed garage, it should be protected from dust, moisture and contaminants.

8. Corrosion control in thermoelectric plants

Electricity is a key element in ensuring economic growth and social development of a country. Many conventional power plants in recent years are being installed in combined cycle power plants, also called cogeneration. The latter, simultaneously generate electricity and / or mechanical power and useful heat, sometimes using thermal energy sources that are lost in conventional plants.

A power station is a thermoelectric energy conversion system, starting with the chemical energy of fuel that during combustion is converted into heat energy accumulated in the steam. This thermal energy generates mechanical energy from the hot steam, which expands in a turbine, turning on electricity in the generator. In this process of low energy thermal efficiency is lost in the hot gases that escape through the chimney and the cooling steam in the condenser.

Electricity generating plants burn fossil fuels such as coal, fuel oil and natural gas. These fuels containing as minor components sulfur compounds (S), nitrogen (N), vanadium (V) and chloride (Cl^-). These are corrosive chemicals attacking the metal infrastructure; and polluting the environment by becoming acid gas emissions, also affecting the health of the population.

The three central equipment of a thermoelectric plant are the boiler, which converts the water into steam, the steam turbine to whom the pressure imparts a rotary motion and the condenser that condenses the vapor released by the turbine and the condensed water is returned to the boiler as feed water. The turbine itself transmits rotary motion to the generator of electricity, which will be distributed to industrial, commercial and homes in cities.

Corrosion in steam plant equipment occurs in two parts of the boiler: on the water side and the steam side, with the fire temperature up to 700 ° C, depending on the type, size and capacity of the boiler. The boiler feedwater must be treated to eliminate the corrosive components: salts such as chlorides and sulfates dissolved oxygen (DO); silicates and carbonates, producing calcareous scale on the boiler walls, regarded as precursors for the formation of corrosion under deposits. The water is softened by eliminating salts and treated to remove oxygen; the pH is controlled by addition of alkaline phosphate to reach a pH range of 10 to 11, and inhibitors are added to the feedwater to prevent corrosion.

Figure 14. Stress corrosion cracking, in a combustion gases pipe of a thermoelectric station.

The flue gases and ash solid particles reach temperatures up to 1000 to 1200 °C, impinging on the outer surface of the boiler water tubes and preheater, creating an atmosphere for aggressive chemical corrosion. The damaged tubes lose its thickness generating metal corrosion products; they often are fractured, suffering a stress corrosion due to the combined effects of mechanical stress and corrosion (Figure 14). Since the tubes lose steam and pressure, the operation of the plant is interrupted and the tubes or its sections should be changed incurring severe economic losses. For example, in the United States has been concluded that the costs of electricity are more affected by corrosion than any other factor, contributing 10% of the cost of energy produced.

A study reveals that in 1991 there were more than 1250 days lost in nuclear plants operating in the United States, due to failure by corrosion, which represented an economic loss of $ 250.000 per day. Such statistics indicate that the power generation industry needs to obtain a balance between cost and methods for controlling effectively corrosion in their plants. It is sometimes advisable to add additives to the fuel, for example, magnesium oxide which prevent the deposition of the molten salts on the boiler tubes. Corrosion occurs also in the combustion air preheater, by sulphurous gases which react with condense and form sulfuric acid. Metal components of the turbine rotor: disks and blades suffer from corrosion by salts, alkalis and solid particles entrained in the vapor. In these cases, it is common to observe the phenomena of erosion-corrosion, pitting and stress corrosion fracture; their damage can be ameliorated through a strict quality control of boiler water and steam.

Efficient maintenance and corrosion control in a power plant is based on the following:

- Operation according to mechanical and thermal regime, indicated by the designer and builder of the plant;
- Correct treatment of fuel, water and steam;
- Chemical cleaning of the surfaces in contact with water and steam, using acidic solutions containing corrosion inhibitors, passivating ammoniacal solutions and solutions;
- Mechanical cleaning of surfaces covered with deposits (deposits), using alkaline solutions and water under pressure;
- Perform an optimum selection of the materials of construction for the components of the plant, including those suitable as protective coatings.
- The installation of online monitoring of corrosion in critical plant areas will be one of the most effective actions to control corrosion. In addition, it is recommended same use and document to use corrosion expert system software and materials databases for the analysis of the materials corrosion behavior.

Corrosion in power plants can be controlled by applying the knowledge, methods, standards and materials, based on corrosion engineering and technology.

9. Corrosion in geothermal environments

The development of alternative energy sources represents one of the most attractive challenges for engineering. There are several types of renewable energies already in operation, such as wind, solar and geothermal. Geothermal environments can lead to aggressive environments, e.g. the geothermal field of "Cerro Prieto", located in Baja California, Mexico.

The physical and chemical properties of the vapor at "Cerro Prieto" make it an aggressive environment for almost any type of material: metal, plastic, wood, fiberglass or concrete. The typical chemical composition of a geothermal brine, is shown in Table 4. Many engineering materials are present as components of the infrastructure and field equipment, required for the steam separation, purification and posterior operations for the generation of electricity. This entire infrastructure is a costly investment and therefore, failure or stoppage of one of them, means economic losses, regardless of how vital it is to maintain constant production of much-needed electricity.

Figure 15. Corrosion in concrete structures used to separate steam from water and to operate steam silencers.

In the process of the geothermal fluid exploitation, corrosion of metal structures occurs from the wells drilling operation, where the drilling mud used, causes corrosion of pumping and piping equipment. Subsequently, when the wells pipes are in contact with the steam, they can also suffer from corrosion-erosion problems, where the corrosive agent is hydrogen sulfide. Steam separators and the pipes are exposed to problems of fouling and localized corrosion due to the presence of aggressive components such as H_2S and chloride ions (Cl^-), present in the wells fluid. These agents lead to the deterioration of reinforced concrete foundations supporting steel pipes, or other concrete structures used to separate steam from water and to operate steam silencers. The reinforced concrete deterioration due to steel corrosion in this aggressive environment, and the steam pressure mechanical forces lead to concrete damage with formation of cracks and fractures.

Component	Na^+	K^+	Mg^{2+}	Ca^{2+}	Cl^-	$SO_4{}^{2-}$	SiO_2	HCO^{3-}
Ppm, mg/kg	6429	1176	18.6	347	11735	15	1133	303

Table 4. Typical chemical composition of typical "Cerro Prieto" geothermal brine

In the power plants, the observed corrosion affects components of the steam turbines, condensers and pipelines, and also the cooling towers and concrete structures inside and outside the building that houses the plant. In these cases, the effects of corrosive attack appears in the form of localized corrosion in metal walls and gas piping) or as corrosion fatigue or stress corrosion, caused by cyclic mechanical forces or residual stresses, in turbines and other metal equipment. Table 5 shows a list of equipment and materials used for construction, which are part of the infrastructure of a geothermal power (Valdez, B. et al., 1999, 2008)

Equipment	Materials
Pipelines	Concrete, steel
Vertical and centrifugal pumps	Steel, copper alloys
Valves	Steel
Flanges and fits	Steel
Silencers	Concrete, steel, FRGP
Brine canals	Reinforced concrete
Evaporation ponds	Plastics
Control and safety instruments	Metals and plastics

Table 5. Equipment and materials used to build infrastructure in a geothermal field

The combination of an aerated moist environment with the presence of hydrogen sulfide gas (H_2S) dissolved in water provides a very aggressive medium (Figure 16), which promotes the corrosion of metals and alloys, such as CS and SS. The presence of dust, from the geothermal field and condensation cycles favor the failure of protective coatings applied to steel, so that developed corrosion leads to constant repairs and maintenance of metal installations: pipes, machinery, cooling towers, vehicles, tools, fences, warehouses, etc.

Cooling towers constructed of wood, steel and fiberglass in the presence of flowing and stagnant water and air currents (induced to complete cooling fans), suffer a serious deterioration of the steel by corrosion and biodeterioration, involving a variety of microorganisms. The timber is subjected to oxygen delignification under the effect of colonies of fungi and algae, as well as fiberglass reinforced polyester screens, which deteriorate due to colonies of aerobic and anaerobic bacteria e.g. sulfate reducers.

Furthermore, carbon steels corrode in the form of delamination due to sulfate reduction processes which induce the oxidation of iron, while the SS nails and screws undergoes localized corrosion, forming pits (Figure 17)

Figure 16. A humid corrosive environment in a geothermal field caused by steam and gases emission.

The deterioration by microorganisms capable of living in these conditions is one of the processes that have provided more information to the study of corrosion induced by microorganisms. In "Cerro Prieto", for example, have been isolated and studied various bacteria capable of growing even at temperatures of 70 ° C under conditions of low nutrient concentrations, while in the geothermal field of "Azufres" bacteria have been isolated to survive at temperatures of 105 °C and pressures of downhole (Figure 18).

Figure 17. Corrosion in a cooling tower of a geothermal power plant.

Figure 18. Biodeterioration of polyester polymeric matrix in a fiberglass screen exposed at geothermal temperatures.

10. Corrosion in the paper industry

Corrosion of the infrastructure used in the pulping and paper industry, is another serious problem for corrosion specialists. The wide experience, gathered from cases of corrosion in the various infrastructure components of the paper industry, has provided an extensive literature on mechanisms, types and control of corrosion in this environment.

In the early 60's of last century, when the continuous digester process was adopted, the paper industry had limited knowledge about caustic embrittlement. Currently, it is known that the digesters are subjected to caustic levels and temperatures too close to the fracture caustic range where the total relieves of stresses in the material are essential. To elucidate the mechanism of this phenomenon, it was necessary to conduct serious investigations, which subsequently provide solutions to the problem of corrosion and caustic embrittlement. Technology in the paper industry has evolved over the last forty years and in parallel we can talk about the solution of corrosion problems in different parts of its infrastructure. Components with high failure rate due to corrosion are those built of bronze, SS, cast iron. Corrosion occurs in the papermaking machinery, where the white water equipment is subjected to an aggressive environment. The metal surfaces are exposed to immersion in this water; to steam that promotes the formation of cracks, which favor the deposit of pulp and other compounds. CS undergoes rapid uniform corrosion, while the copper alloys and SS (austenitic UNS S30400 L: 18% Cr8% Ni, UNS S31600 L: 16% Cr10% Ni 2% Mo) develop localized pitting corrosion. In the mill bleach plants the pulp equipment has traditionally been made of SS which has good general corrosion resistance and weldability. The use of chlorine gas (Cl_2) and oxygen in the bleach plant and pulp bleaching, favors a very aggressive oxidant and SS, as type 317 L (18% Cr14% Ni3.5% Mo). However, in the last 25 years the environment in these plants has become much more corrosive due to the wash systems employed for the paper pulp, which increased the emission of oxidizing and corrosive gases; so type "317 L" SS is not resistant and has a shorter service life. Many mills in the paper industry have opted for the use of high-alloy SS, nickel (Ni) and titanium (Ti), for better corrosion resistance in these particular environments. In general, SS exposed to corrosive environment of bleach plants are benefited by the share of chromium, nickel and molybdenum as alloying elements, which increase their resistance to the initiation of pitting and crevice corrosion. The addition of nitrogen (N) increases its resistance to pitting corrosion, particularly when it contains molybdenum (Mo). Furthermore, to avoid waste of elements such as carbon (C), where a concentration greater than 0.03%, can cause sensitization at affected by heat areas in the solder, causing the SS to be less resistant to corrosion. Other waste elements, such as phosphorus (P) and sulfur (S) can cause fractures in the hot steel, formed in the metal welding area. The corrosive environment of bleach plants contain residual oxidants such as chlorine (Cl_2) and chlorine dioxide (ClO_2), these are added to resists the effects of temperature and acidity, maintaining a very aggressive environment.

Corrosion also occurs in the pulping liquor facilities by sulfites, chemical recovery boilers, suction rolls and Kraft pulping liquors. The Kraft process is the method of producing pulp or cellulose paste, to extract the wood fibers, necessary for the manufacture of paper.

The process involves the use of sodium hydroxide (NaOH) and sodium sulfite (Na_2SO_3) to extract the lignin from wood fibers, using large high pressure digesters. High strength is obtained in the fiber and methods for recovery of chemicals explain the popularity of the Kraft process. The black liquor separated, is concentrated by evaporation and burned in a recovery boiler to generate high pressure steam, which can be used for the plant steam requirements for the production of electricity. The inorganic portion of the liquor is used to regenerate sodium hydroxide and sodium sulfite, necessary for pulping. Corrosion of metals in the facilities used in this process may occur during the acid pickling operation for the removal of carbonate incrustations on the walls and black liquor pipe heaters. It has been found that SS 304 L presents fracture failure and stress corrosion. In the recovery processes of chemical reagents, known as stage re alkalinization, metals can fail due to caustic embrittlement or corrosion-erosion under conditions of turbulent flow. Corrosion also occurs in the equipment used for mechanical pulping, such as stress corrosion cracking, crevice corrosion, cavitation and corrosion-friction.

Author details

B. Valdez[1], M. Schorr[1], R. Zlatev[1], M. Carrillo[1], M. Stoytcheva[1], L. Alvarez[1], A. Eliezer[2] and N. Rosas[3]

1 Instituto de Ingeniería, Departamento de Materiales, Minerales y Corrosión, Universidad Autónoma de Baja California, Mexicali, Baja California, México

2 Sami Shamoon College of Engineering Corrosion Research Center, Ber Sheva, Israel

3 Unversidad Politécnica de Baja California, Mexicali, Baja California, México

References

[1] Acuña, N., Valdez, B., Schorr, M., Hernández-Duque, G., Effect, of., Marine, Biofilm., on, Fatigue., Resistance, of., an, Austenitic., & Stainless, Steel. Corrosion Reviews, United Kingdom (2004)., 22(2), 101-114.

[2] Carrillo, M., Valdez, B., Schorr, M., Vargas, L., Álvarez, L., Zlatev, R., Stoytcheva, M., In-vitro, Actinomyces., israelii's, biofilm., development, on. I. U. D., copper, surfaces., & Contraception, Vol. (2010). (3), 261-264.

[3] Carrillo Irene, Valdez Benjamin, Zlatev Roumen, Stoycheva Margarita, Schorr Michael, and Carrillo Monica, Corrosion Inhibition of the Galvanic Couple Copper-Carbon Steel in Reverse Osmosis Water, Research Article, Hindawi Publishing Corporation, International Journal of Corrosion, Volume(2011). Article ID 856415

[4] 2010, Garcia, A., Valdez, B., Schorr, M., Zlatev, R., Eliezer, A., Haddad, J., Assessment, of., marine, , fluvial, corrosion., of, steel., aluminium, Journal., of, Marine., Engineering, , & Technology, Vol. (18), 3-9.

[5] Garcia Inzunza Ramses, Benjamin Valdez, Margarita Kharshan,Alla Furman, and Michael Schorr, ((2012). Interesting behavior of Pachycormus discolor Leaves Ethanol Extract as a Corrosion Inhibitor of Carbon Steel in 1M HCl. A preliminary study. Research Article, Hindawi Publishing Corporation, International Journal of Corrosion, Article ID 980654, 8 , 2012, 2012.

[6] Lopez, B. G., Valdez, B., , S., Zlatev, R., , K., Flores, J., , P., Carrillo, M., , B., Schorr, M., & , W. Corrosion of metals at indoor conditions in the electronics manufacturing industry. Anti-Corrosion Methods and MaterialsUnited Kingdom, N0. 6, Noviembre (2007). , 54, 354 EOF-359 EOF.

[7] Lopez, B. G., Valdez, B., , S., Schorr, M., , W., Rosas, N., , G., Tiznado, H., , V., Soto, G., & , H. Influence of climate factors on copper corrosion in electronic equipment and devices. Anti-Corrosion Methods and MaterialsUnited Kingdom, N0. 3, (2010). , 57, 148 EOF-152 EOF.

[8] Lopez, Gustavo, Hugo Tiznado, Gerardo Soto Herrera, Wencel De la Cruz, Benjamin Valdez, Miguel Schorr, Zlatev Roumen,(2011). Use of AES in corrosion of copper connectors of electronic devices and equipments in arid and marine environments. Anti-Corrosion Methods and MaterialsIss: 6, , 58, 331-336.

[9] Lopez, Badilla Gustavo, Benjamin Valdez Salas and Michael Schorr Wiener, Analysis of Corrosion in Steel Cans in the Seafood Industry on the Gulf of California, Materials Performance, Vol.April (2012). (4), 52-57.

[10] Navarrete, M., Ballesteros, M., Sánchez, J., Valdez, B., Hernández, G., Biocorrosion, in. a., geothermal, power., plant, Materials., & Performance, April. (1999). USA., 38, 52-56.

[11] Schorr, M., Valdez, B., Zlatev, R., Stoytcheva, M., Erosion, Corrosion., in, Phosphoric., Acid, Production., Materials, Performance., & Jan, . (2010). USA, 50(1), 56-59.

[12] Stoytcheva, M., Valdez, B., Zlatev, R., Schorr, M., Carrillo, M., Velkova, Z., Microbially, Induced., Corrosion, in., The, Mineral., Processing, Industry., Advanced, Materials., & Research, . (2010). Trans. Tech publications, Switzerland, 95, 73-76.

[13] Raicho Raichev, Lucien Veleva y Benjamín Valdez, Corrosión de metales y degradación de materiales.Principios y prácticas de laboratorio. Editorial UABC, 978-6-07775-307-0(2009). pp

[14] Santillan, S. N., Valdez, S. B., Schorr, W. M., Martinez, R. A., Colton, S. J., Corrosion, of., the, heat., affected, zone., of, stainless., steel, weldments., Anti-Corrosion, Methods., Materials, United., & Kingdom, Vol. (2010). (4), 180 EOF-184 EOF.

[15] Schorr, M., Valdez, B., Zlatev, R., Stoytheva, M., Santillan, N., Phosphate, Ore., Processing, For., Phosphoric, Acid., Production, Classical., And, Novel., Technology, Mineral., Processing, , Extractive, Metallurgy., & Vol, . (2010). (3), 125-129.

[16] Valdez, B., Guillermo Hernandez-Duque, Corrosion control in heavy-duty diesel engine cooling systems, CORROSION REVIEWS Vol.Nos. 2-4, (1995). , 245-260.

[17] Valdéz, Salas. B., Miguel, Beltrán., Rioseco, L., Rosas, N., Sampedro, J. A., Hernandez, G., & Quintero, M. Corrosion control in cooling towers of geothermoelectric power plants. Corrosion Reviews, (1996). England., 14, 237-252.

[18] Valdez, B., Rosas, N., Sampedro, J., Quintero, M., Vivero, J., Hernández, G., Corrosion, of., reinforced, concrete., of, the., Rio, Colorado., Tijuana, aqueduct., Materials, Performance., & May, . (1999). USA., 38, 80-82.

[19] Valdez, B., Rosas, N., Sampedro, J., Quintero, M., Vivero, J., Influence, of., elemental, sulphur., on, corrosion., of, carbon., steel, in., geothermal, environments., Corrosion, Reviews., & Vol, . Nos. 3- 4, October (1999). England, 167-180.

[20] Valdez, S. B., Zlatev, R., , K., Schorr, M., , W., Rosas, N., , G., Ts, Dobrev. M., Monev, I., Krastev, Rapid., method, for., corrosion, protection., determination, of. V. C. I., Films-Corrosion, Anti., Methods, , Materials, United., & Kingdom, Vol. Noviembre (2006). (6), 362-366.

[21] Valdez, B., Carrillo, M., Zlatev, R., Stoytcheva, M., Schorr, M., Cobo, J., Perez, T., & Bastidas, J. M. Influence of Actinomyces israelii biofilm on the corrosion behaviour of copper IUD, Anti-Corrosion Methods and Materials, United Kingdom, N0. 2, 55-59, (2008)., 55

[22] Valdez, B., Schorr, M., Quintero, M., Carrillo, M., Zlatev, R., Stoytcheva, M., Ocampo, J., Corrosion, , scaling, at., Cerro, Prieto., Geothermal, Field., Anti-Corrosion, Methods., Materials, United., & Kingdom, Vol. N0. 1, (2009)., 28 EOF-34 EOF.

[23] Valdez, B., Schorr, M., Corrosion, Control., in, The., Desalination, Industry., Advanced, Materials., & Research, . (2010). Trans. Tech publications, Switzerland, 95, 29-32.

[24] Valdez, B., Schorr, M., Quintero, M., García, R., Rosas, N., The, effect., of, climate., change, on., the, durability., of, engineering., materials, in., the, hydraulic., & infrastructure, . An overview. Corrosion Engineering Science and Technology(2010). , 45(1), 34-41.

[25] Valdez, B., Schorr, M., So, A., Eliezer, A., Liquefied, Natural., Gas, Regasification., Plants, Materials., Corrosion, M. A. T. E. R. I. A. L. S. P. E. R. F. O. R. M. A. N. C. E., & Vol, . December (2011). (12), 64-68.

[26] Vargas, O. L., Valdez, S. B., Veleva, M. L., Zlatev, K. R., Schorr, W. M., Terrazas, G. J., Corrosion, of., silver, at., indoor, conditions., of, assembly., processes, in., the, microelectronics., industry-Corrosion, Anti., Methods, , Materials, United., & Kingdom, Vol. N0. 4, (2009)., 218 EOF-225 EOF.

[27] Veleva, L., Valdez, B., López, G., Vargas, L., Flores, J., Atmospheric, Corrosion., of-Electronics, Electro., Metals, in., Urban-Desert, Indoor., & Environment, . Corrosion of Electro-Electronics Metals in Urban-Desert Indoor Environment. Corrosion Engineering Science and Technology(2008)., 43(2), 149-155.

Chapter 3

Wear and Corrosion Behavior in Different Materials

N. A. de Sánchez, H. E. Jaramillo, H. Riascos,
G. Terán, C. Tovar, G. Bejarano G., B. E. Villamil,
J. Portocarrero, G. Zambrano and P. Prieto

Additional information is available at the end of the chapter

http://dx.doi.org/10.5772/52986

1. Introduction

This chapter presents three articles with results, about our research, relationships with the behavior corrosion resistant in different materials of engineer. The first research is about TiC monolayers and Ti/TiN/TiC/TiCN multilayers were deposited onto AISI 4340 steel substrates at 300° C and $4.0x10^{-2}$ mbar nitrogen/methane pressures by using a Pulsed-Laser Deposition technique. A Nd: YAG laser (1064 nm, 500 mJ and 7 ns), with a repetition rate of 10 Hz was used. Here we analyzed the difference in structure and morphology for single layers and multilayers structures deposited on stainless steel substrates. AFM analysis presented different morphologies; showing, that the single layer had an average grain size of 0.087 μm; while the multilayers exhibited grain sizes of 0.045 μm. Coating thicknesses were 1 μm, approximately, and monolayer average roughness was 0.21 μm; while a value of 0.15 μm was measured for multilayers. Under a 60 Kg maximum load applied in tension to evaluate the adhesion to the substrate, no detachment of the films was presented. Multilayers evidenced better impact resistance as compared with TiC single layer; this result is considered, bearing in mind that in multilayers propagation of fissures is slower, because the presence of layer inter-phases, lead to fissures strays in other directions. Slight corrosion specks are present; but mass loss was around 16 mg. in multilayers a value was lower than for the TiC single layer that was near 43 mg. Homogeneity, grain size, fracture resistance, corrosion resistance, and adhesion of the multilayers are suitable for mechanical applications of these types of coatings as shown in mechanical measurements. These results indicate that for engineering applications under corrosive environments, the use of this type of multilayers coatings on AISI 4340 steel, are highly recommended.

In the second research we present results of corrosion test analysis on 304-type stainless steel welded samples. The samples were welded by SMAW, GMAW, and GTAW techniques. Corrosion tests were carried out under normal working conditions. The samples were installed on the ventilation extractor at a sugar mill, with a pull inducement to facilitate the rise of gases containing concentrations of smoke and sulfur and sulfide vapors, which generate a low PH within the tower. The samples were removed from the ventilator vents at 15-day programmed intervals, until completing 16 samples, for the purpose of studying the advancement of corrosion dependent of exposition time. Filler welt morphologies were analyzed by using Scanning Electron Microscopy SEM; the interface between the filler metal and base metal was studied by means of an image analyzer. SEM microscopy revealed that samples treated with the GMAW welding process presented the phenomenon of segregation upon the solder welt. Through EDX analysis we found the formation of carbides, which explains the presence of the segregation phenomenon. Due to the loss of chromium at the solder welt, there was the generation of a greater corrosion zone in these simples treated with GMAW processes compared to simples treated with SMAW and GTAW welding techniques. The migration of carbon to zones of greater segregation favored the carbide formation, which cause diminished corrosion resistance. Additionally, given the chemical dissimilarity between the segregated and non-segregated zones, a galvanic pair was generated that increased the phenomenon.

And the final we show WC/W coatings were deposited by reactive magnetron sputtering using 40%, 60% and 80% methane CH4 in the gas mixture. The bilayers were grown on to AISI 420 stainless-steel substrates in order to study the wear and corrosion behavior. Before growing the bilayers, one Ti monolayer was grown to improve the adherence of the coatings to the substrate. The wear resistance and the friction coefficient of the coatings were determined using a pin-on-disk tribometer. All coatings had a friction coefficient of about 0.5. The measured weight lost of the bilayers from each probe allowed the qualitative analysis of wear behavior all coatings. The bilayers grown with 80% methane showed the best abrasive wear resistance and adhesion without failure through the coating in the wear track for dry pin-on-disk sliding. Electrochemical corrosion test showed that the bilayers grown with 80% methane were more resistant to corrosion than the ones uncoated.

2. Preface

The need to improve the physical/mechanical properties of materials has led to the development of new hard and super hard coatings with hardness over 40 GPa as compared with the hardness of diamonds, (96±5 GPa) [1]. Some examples of these coatings are: Carbon Nitrate [2,3,4], Nitrate super networks [5], and Oxide super networks [6], and high wear resistance [7].

Hard coatings that are mostly resistant to wear and corrosion have been commercially utilized to increase the useful life of a number of industrial elements, such as: cutting tools, gears, bearings, and industrial machinery components [8]. Applications include protection coatings, thermal barriers, optical applications, bio-medicine, semiconductors, and decorative uses. In the last decades, intensive work has been carried out on hard coatings by different research groups. With hard coatings it is possible to improve the surface properties of materials or those properties that are dependent of the surface, as are: hardness; corrosion, fatigue, and wear resistance. [8] These properties, together with the electrical, magnetic, and optical properties are not only of scientific interest, but also of technological interest, being that in both mechanical and biological systems it is necessary to improve the useful life of elements; save energy; and improve their efficiency and reliability.

Metallic materials, especially steel, upon which hard coatings are applied, present high surface hardness, low friction coefficient, high resistance to wear, high resistance to fatigue and corrosion, and dimensional stability at working temperatures bellow 500 °C [4]. The aforementioned characteristics are dependent of several factors, like the type of material to be coated, the kind of coating, and compatibility and adhesion of the hard coating upon the material to be coated. In the specific case of pulsed-laser ablation technique, certain parameters should be observed; such as: the chamber's vacuum pressure, pulse duration, rate of repetition, temperature of the substrate, gas ratio. In film growth we must bear in mind the crystalline structure, the film's homogeneity, density, thickness, and its chemical and morphological composition.

Recently, there has been a great deal of attention to the production of multilayers, mainly multilayers of ceramic/ceramic transition metal nitrates, each of the layers in the scale of nanometers, due to the possibility of obtaining super hardness [9,10,11]. There is improvement in the tribologic properties of metal/metal nitrates or in the multilayers of oxides; this area has attracted the interest of research groups [12]. In recent studies, J.P. Tu et al, [13]. analyzed Ti/TiN multilayers with a soft compositional transition, which they produced with computer-controlled gas flow. The results revealed that both the interfaces, as well as the period modulation of the multilayers play an important role in improving mechanical properties [14,15]. This work compares the structure, morphology, grain size, fracture resistance, wear resistance, and corrosion resistance of multilayers as opposed to monolayers.

In industrial processes, one of the most widely used materials is stainless steel, given its excellent anticorrosion, mechanical, and aesthetic properties. But when stainless steel is attacked by acidic agents under certain concentration parameters, its behavior as an anticorrosive material may not be the best; and steel can be attacked with great severity.

Corrosion is defined as the deterioration of material as a result of chemical attacks within its environment [20]; corrosion refers specifically to any process involving the deterioration or degradation of metal components [21]. More specifically, it can be defined as a chemical or electrochemical reaction between a material – normally a metal – and its environment, producing deterioration of the material and its properties [22]. As much as corrosion is originated from a chemical reaction, the velocity at which it will take place will depend, in some measure, on temperature, salinity of the fluid, and properties of metals in question [23].

Different types of corrosion can appear on stainless steel, depending on the aggressor medium and surface conditions, roughness, chipping and residual efforts. The types of corrosion that can appear are inter-granular corrosion, corrosion by chipping, corrosion by efforts, galvanic corrosion [24] etc. Bearing in mind the aforementioned, we present the study of deterioration of austenitic 304 stainless steel joined through welding with SMAW, GTAW, and GMAW [25] processes and exposed to an industrial environment, namely, the sulfitation tower in a sugar mill, subjected to sulfur gases.

It is well established that the mechanical, tribological and thermal properties of many metals can be significantly improved by the controlled incorporation of interstitial elements, such as boron, nitrogen and carbon. Although W and WC coatings have been used for a number of applications [31], there seems to be a lack of knowledge about coatings with carbon concentrations [32] at the intermediate range below that of stoichiometric WC. Tungsten carbide, silicon carbide and silicon nitride are predicted to be among the most abrasion-resistant tool materials [33] and are known to be effective in resisting erosion [34]. Dimigen and co-workers developed tungsten metal doped DLC (W:C-H) coatings with W/C ratios of around 0.1 [35,36].These have attracted considerable attention, owing to their high wear resistance and low friction coefficient.

One of the major drawbacks of the wear tests is the scarce reproducibility of the experimental data. In citing the results of a UK inter-laboratory project of a single pin-on-disk test on wear-resistant steels worn under fixed test conditions, Almond and Gee [37] reported dispersion in the range of 57.2–75.4% for the wear data and a minimum reproducibility of 37% for the friction coefficient. Considering the repeatability within laboratories, the latter value was in the range of 3–19%. These great dispersions, especially for the wear data, were mainly attributed to the difficulty of properly mating the flat-end pin surface with the disk surface in the different test machines of the laboratories involved in the project.

3. Experimental details mechanical behavior of TiC monolayers and Ti/TiN/TiC/TiCN multilayers on AISI 4340 steel

TiC monolayers and the multilayers analyzed in this work were grown on AISI 4340 steel through the laser ablation technique. An INDI-30 spectra-Physics TM Q-switched Nd: YAG pulsed laser was used with energy at 500 mJ, pulse duration of 7 ns, a wavelength of 1064 nm, and a repetition rate of 10 Hz. The TiC films were grown from a high-purity Ti target in the order of 99.999% in an atmosphere of Ar, N_2, and CH_4 gases. The coatings were grown at a temperature of 300 °C; pressure of 4.0×10^{-2} mbar was kept constant during the growth of all films. The thickness of the films was approximately 1 μm.

The films were characterized by Atomic Force Microscopy (AFM), Electronic Scaning Microscopy (SEM), resistance to wear, resistance to fracture, and resistance to corrosion. This

http://dx.doi.org/10.5772/52986

work analyzed the difference in structure and morphology of the monolayers and multilayers deposited upon a similar type of substrate.

The adherence of the films grown on steel substrates was estimated by way of tests, using normalized Y-type adhesive tape of L-T-90[1] specifications according to procedures described in norm MIL – F – 48616 [2]. For such, the tape was firmly applied over the film, without reaching the borders to keep from tearing the films at inhomogeneous regions that could be present on the contour and, thereby, avoiding the formation of air bubbles. Thereafter, while the substrate was being held with tweezers, the tape was quickly removed at a normal angle to the film surface. All monolayers and multilayers analyzed passed the test with no apparent visual deterioration, thus indicating that the adherence to the substrate is acceptable.

Also, adhesion tests of the films were carried out through the application of tension loads. The substrate with the film was placed over a test tube used for the tension trial. The sample was adhered with Loctite 495 glue, superdner instant adhesive.

Progressive loads were applied 5Kg at a time till reaching 60Kg. Upon applying the maximum 60Kg load, the substrate with the film separated from the glue, but the film did not reveal any detachment after the visual check, with which it is concluded that the adherence of the films is acceptable and superior to the capacity of the glue's adhesion.

The TiC films and the multilayers were exposed to a wear test by using a G99 pin on disc type tribometer. The pin corresponded to the analyzed film and the AISI D2 steel disc with a roughness of 0.71μm. was worked with a velocity of 0.1 m/s. A 30 gr. load was applied for 5 min. The prior conditions remained constant during the analysis of all the films.

The film's resistance to fracture was assessed using the Rockwell C hardness test. As of the analysis of traces left by the indenter, it was found that the multilayers are more resistant to fracture than the TiC monolayers. Adhesion of the films was determined by the tension test; maximum loads of 60 Kg were applied and neither the monolayers nor the multilayers showed signs of substrate detachment.

The monolayers and multilayers were exposed to corrosion-erosion tests, using the equipment shown in figure 1. The specimen were impingement with a slurry composed by a 0.2 pH solution with 30 wt.% of SiO_2 particles with mean size between 212 and 300 μm. The impeller device was a UHMWPE disk at 3500 rpm rotation speed. The surface impact area was 0.05 m^2; the slurry temperature varied between 25 and 28 °C and the test time were one (1) hour.

The mean impact angle was 30 °C but it is worth to notice that the dispersion is great due to several factors as boundary layers and viscosity effects, that deviate the particles as they are reaching the probes surfaces [18].

1 Tape As Per MIL-A_AA-113B

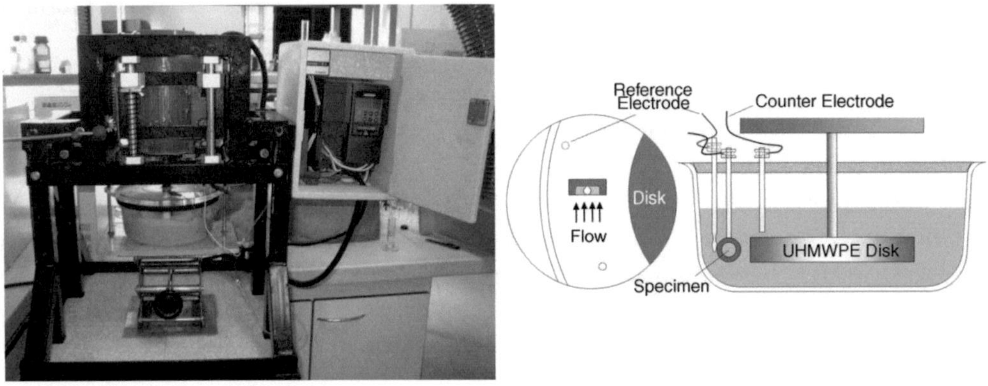

Figure 1. Experimental setup for liquid impingement and corrosion–erosion tests.

4. Results and analysis

All the films underwent AFM analysis to determine the morphology and structure as a function of the substrate. AFM images in figure 2 show the TiC film and the multilayer, revealing, in both cases, high-homogeneity surfaces free of pinholes or agglomerates. The multilayer has very fine and uniform grain size, while the TiC monolayer surface shows somewhat larger grain size and a less homogenous distribution.

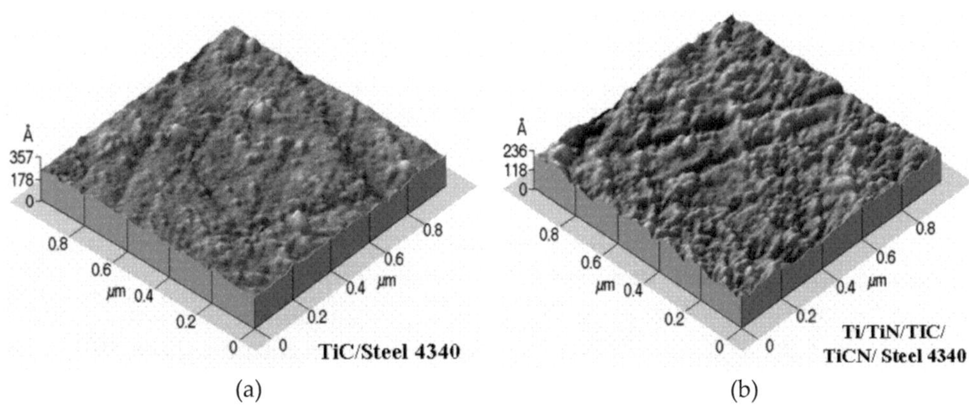

Figure 2. AFM Micrographs of a) TiC and b) multilayers grown on AISI4340 steel.

Two-dimensional AFM micrographs in figure 3 show the differences in grain size for the monolayers and the multilayers: the monolayers had an average grain size of 0.087 μm, while the multilayers exhibited grain sizes of 0.045 μm. Roughness for the monolayers was at 0.21 μm, and for the multilayers it was at 0.15 μm.

http://dx.doi.org/10.5772/52986

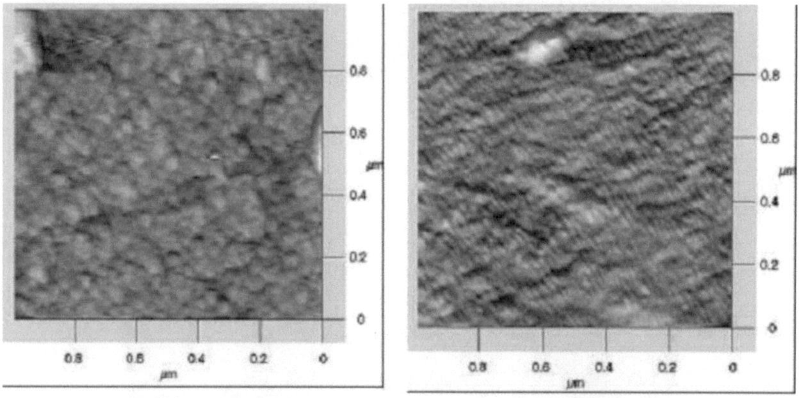

(a) (b)

Figure 3. AFM Micrographs of a) TiC monolayers and of b) Multilayers, where the grain- size difference is observed.

Monolayers and multilayers were exposed to a wear test by using a G99 pin on disc type tribometer. The micrograph of the TiC film in figure 4a shows where there was the most wear on the film; the wear produced was of 0.0008 gr., while the multilayer did not reveal wear at the end of the test according figure 4b.

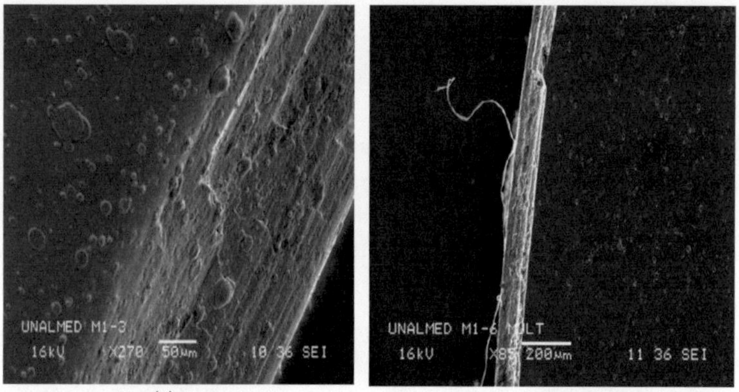

(a) (b)

Figure 4. SEM Micrographs after wear test: a) TiC monolayer, b) multilayer.

The coatings were characterized through the Rockwell C hardness test to qualitatively determine their resistance to fracture. Loads of 60 Kg, 100 Kg, and 150 Kg were applied to the monolayers and the multilayers. Optical photographs were taken of the traces left by the indenter upon the application of the three different loads. Figure 5a shows traces left by the indentations carried out upon the TiC coatings at different loads. Upon observation of microscopic images, radial fractures are detected toward the interior – keeping the radial directionality and without notable detachment of the film.

The micrographs in figure 5b correspond to Rockwell C indentations applied to the multilayers. It was microscopically observed that the plastic deformation around the trace is very slight and that the film did not suffer any detachment. When load is increased, densities of the fissures around the trace diminish in density.

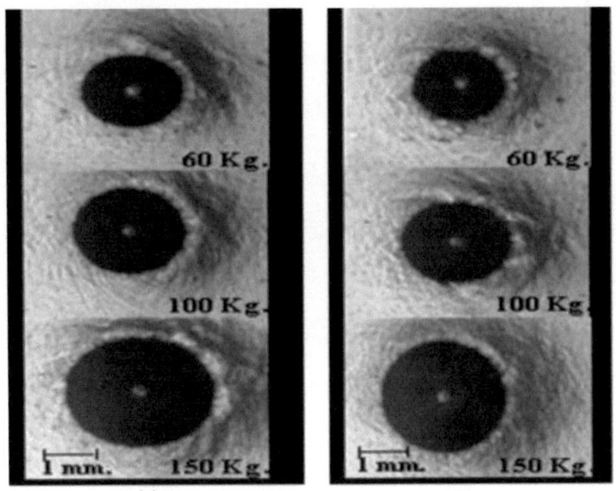

(a) (b)

Figure 5. Images of Rockwell C indentation traces a) upon TiC monolayers, and b) on multilayers, (charges are indicated).

The generation of circumference fissures observed is due to radial tension efforts outside of the indentation zone. As the load is increased, these are submitted to compression efforts. The propagation of fissures with a load of 60 Kg. curve and run parallel to the surface, such is explained by a field of compressive tensions according to Bhowmicks et all [19]. The multilayers showed better impact resistance as compared with the TiC monolayers. This result is considered, bearing in mind that in the multilayers the propagation of fissures is slower, because on the interphase of the coatings the fissures stray in another direction.

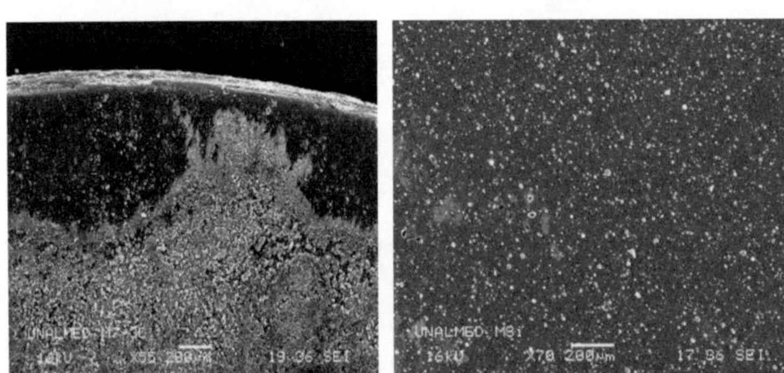

(a) (b)

Figure 6. SEM Micrographs after corrosion test in a saline atmosphere a) TiC film b) multilayer.

As can be seen in figures. 6a and 6b corresponding to SEM micrograph of TiC and multilayer coatings respectively. The TiC films were severe worm out while on the multilayers we observed formation of pits. The TiC mass lost was 0.043 g and the multilayers′ one was 0.016. g. We observed that the monolayers were more vulnerable to corrosion - erosion test.

5. Experimental details corrosion effects on SMAW, GMAW, and GTAW welding processes stainless steel 304

In the development of the welding process, GMAW, GTAW, and SMAW techniques were used with welding equipment. The first arc welding process was applied following international welding norms, using austenitic 304 stainless steel plates [26] with dimensions of 50 x 20 x 4.76 mm. A coated electrode series E308L-16 was used. The time of application of the welding chord was of 1′57″64‴, with power at 38V and current at 85A. The same characteristics, variables and equal working conditions were used for GTAW and GMAW welding processes.

The corrosion test, to which the samples were subjected, is classified as a field test [22] With this test we seek to determine which is the Contributing Metal and Base Metal in the Thermally Affected Zone that is least affected by the aggressive medium, that fundamentally is acidic type like smoke emissions and sulfur vapors, as well as its relationship with the welding process used.

The experimental phase was carried out at a sugar mill, since this type of industry is one of the most affected by corrosion problems. It was determined that the most severe corrosive medium and the highest corrosion levels in the sugar mill for the tests took place in the sulfitation towers, where there are elevated concentrations of sulfur and sulfides smokes and vapors, which generate low PH [27] which is an ideal environment for the study of corrosion in the Contributing Metal and Base Metal in the Thermally Affected Zone, in 300-series stainless steels like the 304.

The samples were installed at the ventilator - extractor damper where draft is induced to facilitate the ascension of gases within the tower, as seen in figure 7a. The samples were removed from the ventilation vents at 15-day programmed intervals until completing a total of six (6) samples for the study of corrosion advance. Figure 7b shows the ventilator extractor where the samples were placed, allowing them to be attacked by the vapors expelled through this place.

The samples were metallographically analyzed through the Leyca Q'Metals' image analyzer. Images were taken using an optical microscope. Through this procedure we were able to establish which of the welding processes yielded the highest deterioration due to corrosion. Then, we obtained Scanning Electron Microscopy (SEM) images to analyze surface morphology.

To obtain optical microscopy images, the samples were attacked with a solution that included: 5 cm3 of nitric acid, 3 cm3 hydrochloric acid, and 2 cm3 of hydrofluoric acid mixed in 20 cm3 of water. The samples were kept submerged for 4 seconds each.

(a) (b)

Figure 7. Photos of the attack-sites of samples: a) Ventilator extractor where samples were placed, b) Ventilator damp where samples were anchored.

6. Results and analysis

Optical microscopy analyses reveal that in samples welded with the GMAW process, a segregation phenomenon occurred on the welt – as shown in figures 8a and 8b. It was also found that in the segregated zones, corrosion was quite high, as observed in figures 9a, 9b, 10a, and 10b.

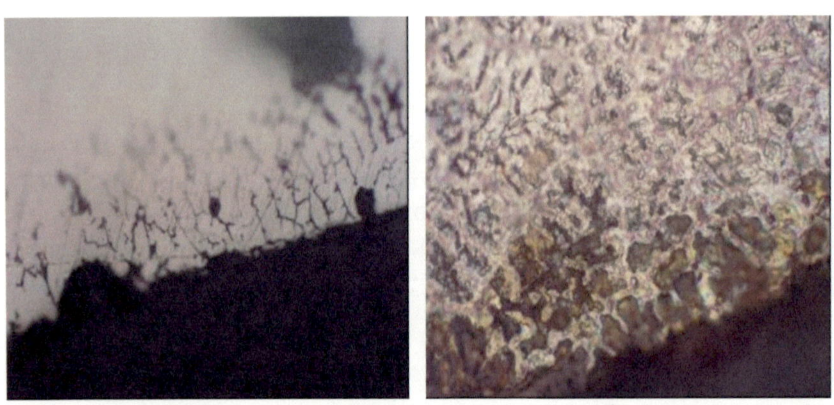

(a) (b)

Figure 8. Optical Microscopy Images of samples with GMAW welding process; removed after 1.5 months: a) Non-attacked sample – X50 welding welt, b) attacked sample, X500 fused zone.

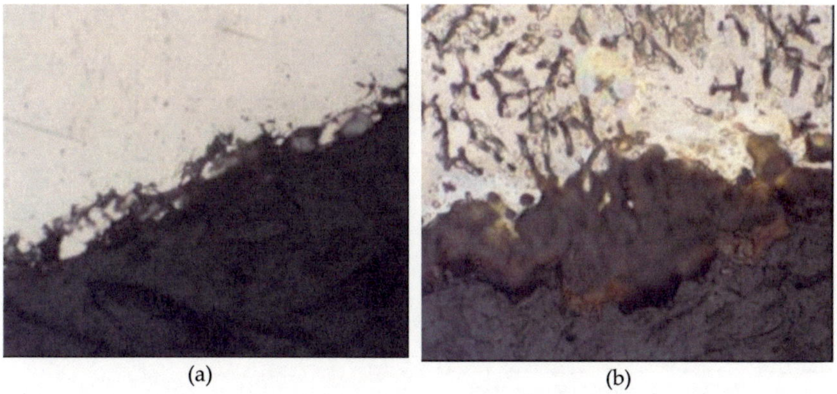

Figure 9. Optical Microscopy Images of samples with SMAW welding process; removed after 1.5 months: a) Non-attacked sample – X500 welding welt, b) attacked sample, X500 fused zone.

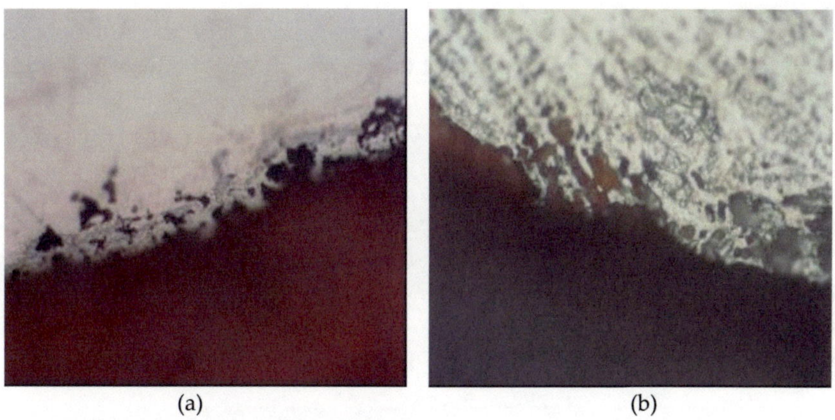

Figure 10. Optical Microscopy Images of samples with GTAW welding process; removed after 1.5 months: a) Non-attacked sample – X500 welding welt, b) Zone defused and attacked with acid.

In order to analyze the segregated zone and establish this zone's morphology and composition, a scanning optical microscopy and EDX microprobe were used, carrying out particular microanalysis on the segregated zone according to figures 11a, 11b, 12a, and 12b and on non-segregated zones as evidenced in figures 13a and 13b. Through the microprobe analysis we found the possible existence of chromium carbides formation in these zones, especially in the GMAW process, which would explain why corrosion was originated on segregated zones where carbon migration to the zone with the greatest segregation took place, with which an increased concentration of carbon can be obtained, increasing the possibility of carbide formation and decreasing resistance to corrosion. In addition, given the chemical dissimilarity between the segregated and non-segregated zones, a galvanic pair is generated, which increases the phenomenon.

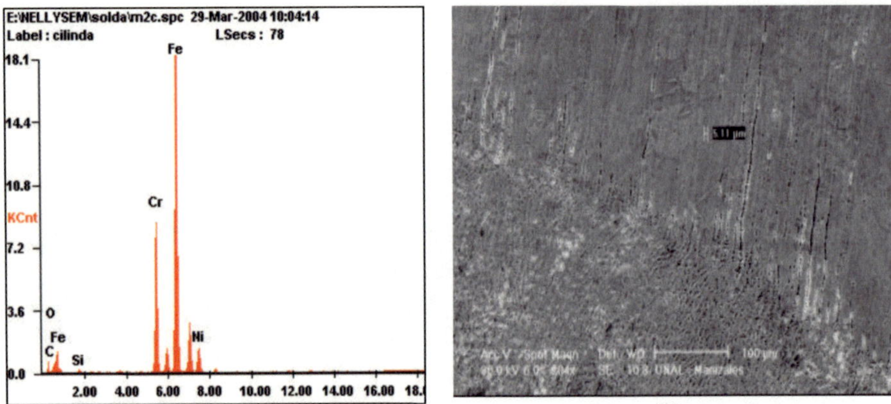

(a) (b)

Figure 11. a) EDX Spectrum, b) SEM image of segregated zone of samples with GTAW processes; removed after 1.5 months.

Once optical microscopy and SEM images of samples welded with the GMAW process were compared and analyzed, it was observed that these were the most affected by inter-granular corrosion, as seen in figures 8, 12, and 13. In the segregated zones within the welt, samples treated with SMAW processes (figure 9) and GTAW processes (figures 10 and 11) corrosion advance was not as significant.

The comparative analysis among the samples revealed, in time, corrosion increase, probably due to the fact that corrosion advances through the segregated zone and presenting a phenomenon of corrosion through chipping [22, 24 and 29] which began as intergranular type corrosion.

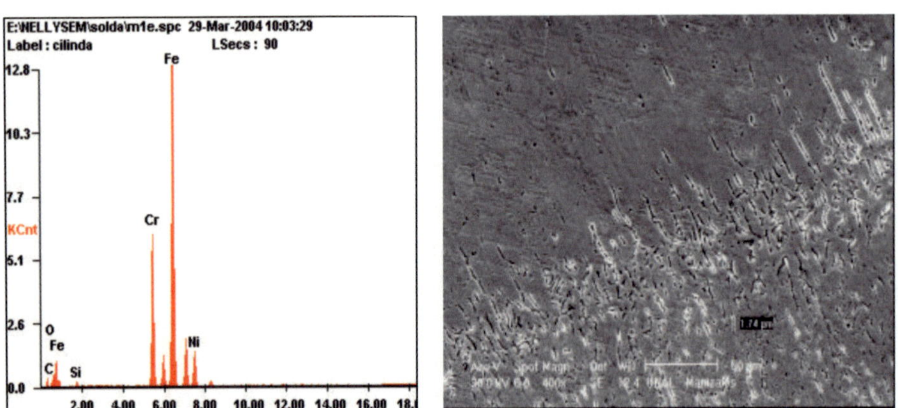

(a) (b)

Figure 12. a) EDX Spectrum, b) SEM image of segregated zone of samples with GMAW processes; removed after 1.5 months.

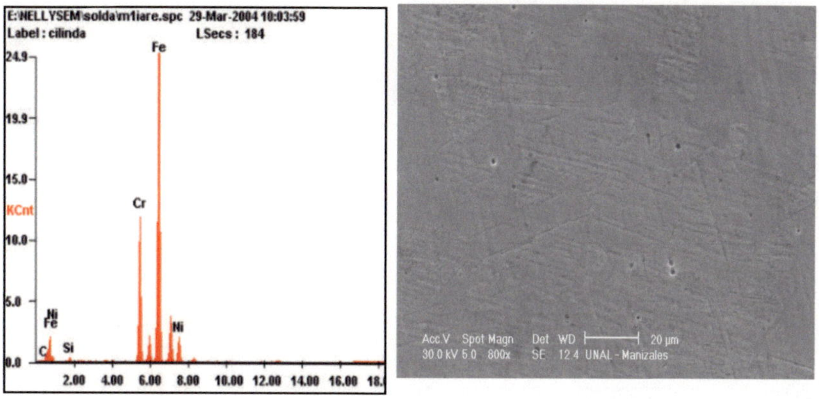

(a) (b)

Figure 13. a) EDX Spectrum, b) SEM image of non-segregated zone of samples with GMAW processes; removed after 1.5 months.

The literature mentions that intergranular corrosion is generated in the thermally affected zone [29], in this research the problem of inter-granular corrosion occurred in the solder welt, especially with the GMAW process, probably due to the segregation phenomenon [11] that could have caused the precipitation of M7C3 or M23C6 chromium carbides [30], given the chemical composition registered in the zones observed in the EDX Spectra.

7. Experimental details wear and corrosion behavior of W/WC bilayers deposited by magnetron sputtering

In this work, bilayers of tungsten W and tungsten carbide WC by r.f. magnetron sputtering reactive technique were deposited on AISI 420 stainless steel of 16 mm diameter and 3.8 mm thick substrates. The used parameters were power density 0.045 W/mm2; working pressure 5 mtorr; gas mixture with 40%, 60% and 80% CH4 balanced with Argon; W-Target 99.99% purity, 100 mm diameter; 70 V bias voltage and 70 mm distance between Target and substrate.

Previous works [38], where the microstructure and composition of WC coatings, as methane content in the gas mixture Ar/CH4 were investigated, showed that when the methane content was less than 10% the structure was principally W2C hexagonal phase [39].For methane contents between 17-23% [40] a structure composed by the cubic WC1-x with a preferred (111) orientation imbibed in an amorphous carbon matrix. A hardness level of about 25 GPa was attained and approximately 40% W in the coating. Above this methane tenors the amorphous carbon phase becomes more evident, this phase has a lower friction coefficient and it is also auto lubricant, and wear resistance is increased.

The steel was characterized by chemical and metallographic analysis; the specimens were heat treated by quenching from 1000 °C into oil and tempered at 200 °C for 1 hour; Rockwell

C hardness and roughness with 0.8 mm sensor shifting tests were carried out. The measured Rockwell C hardness was the mean of 5 tests given the following results, 14 RC before the heat treatment and 27 RC as quenching and tempered.

Before growing the bilayers a thin Ti film was grown to better the adherence between bilayers and steel. The corrosion behavior of the coatings was analyzed by using the potentiostatic test method. To determine the wear resistance the pin on disk test was used with the following parameters:

Parameters	Values
Pin initial weight (g)	0.8813
Uncoated probe initial weight (g)	5.8155
Pin diameter (mm)	6
Applied load (N)	2
Displacement (m)	2500
Angular speed (rpm)	100
Trace wear radii (mm)	3.5

Table 1. Experimental parameters setting used with the pin on disk test.

8. Results and discussion

8.1. Heat treatment of steel substrates

The AISI 420 steel micrographs, before and after the heat treatment HT are presented in figure 14. In figure 14a an austenitic phase is present while in figure 14b after heat treatment, it could be observed a martensitic phase with small carbides. These structures are in accordance with the SAE metals handbook.

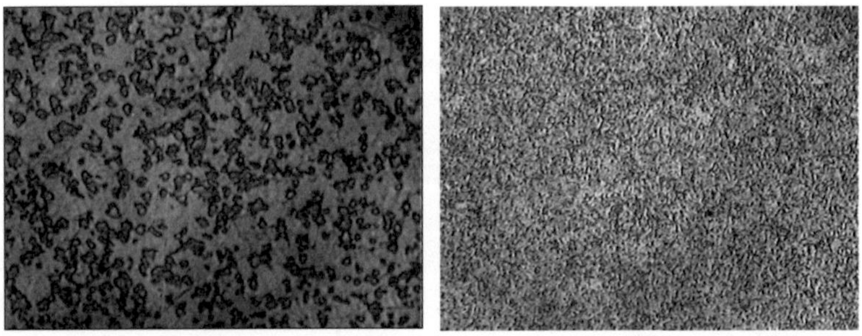

(a) (b)

Figure 14. AISI 420 steel micrographs a) without heat treatment HT and b) after heat treatment HT.

http://dx.doi.org/10.5772/52986

The Rockwell C hardness here reported is the mean of 5 tests given the following results, 14 RC before the heat treatment and 27 RC as quenching and tempered.

8.2. Corrosion behavior of coating system

In figure 15 the polarization curves are presented. The Taffel curves are registered under static conditions and correspond to probes with and without coatings. The corresponding ones to films grown with 60% CH4 and 80% CH4 showed a shifting to lesser current densities indicating a higher corrosion resistant as compared with that of the probe without coating, and the one corresponding to 40% CH4.

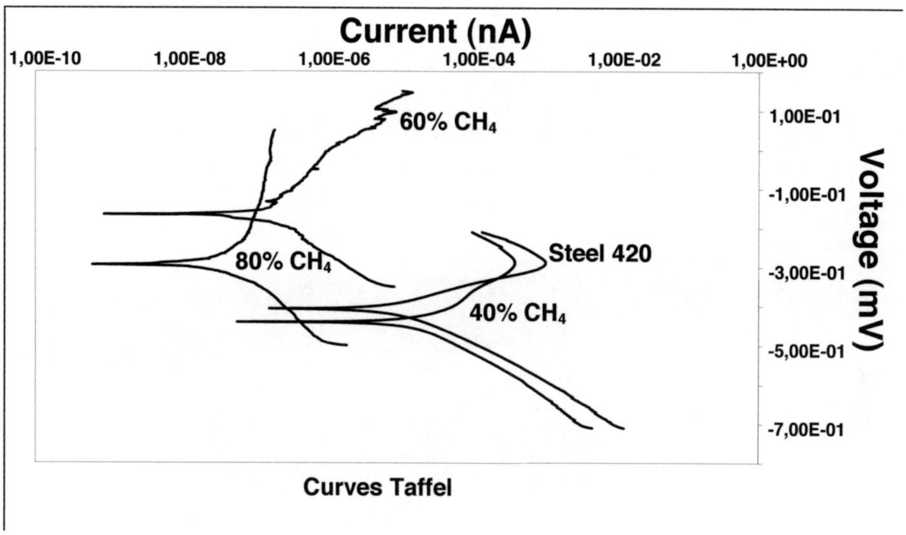

Figure 15. Polarization Curves of the different steel substrates without and with bilayers.

These two last curves behave similarly. It is possible that the low coating adherence were a consequence of the peeling of the coating associated with an anodic current peak presence as can be seen in figure 15, further visual analysis allows supposing that the presence of located corrosion due perhaps to weak mechanical stability of the coating, dissolves the substrate under the coating with an inevitable peeling. The 60% and 80% CH4 curves exhibit a tape increase of the anodic curve and corrosion intensity decrease indicating that the system is under anodic control. Figure 16 shows an AISI 420 steel without coating. A homogeneous attack over the entire surface is observed.

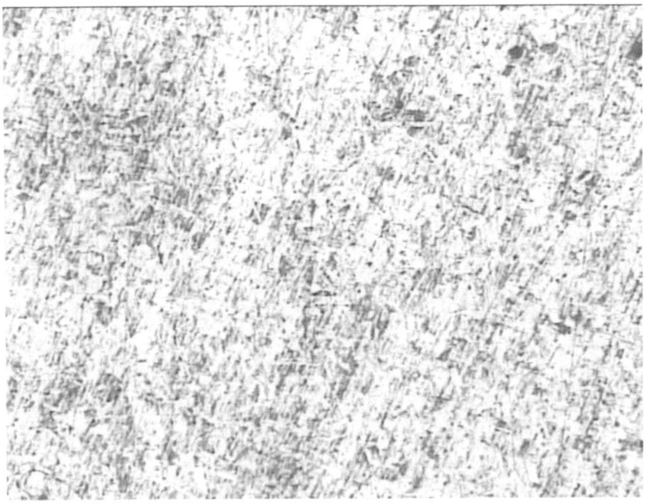

Figure 16. AISI 420 steel micrograph attacked by electrochemically corrosion.

In Figure 17 the surfaces of the bilayers grown at 40% 60% y 80% CH4 after the corrosion test are presented. The one at 40% CH4 were peeled, while the other two presented a kind of pitting corrosion.

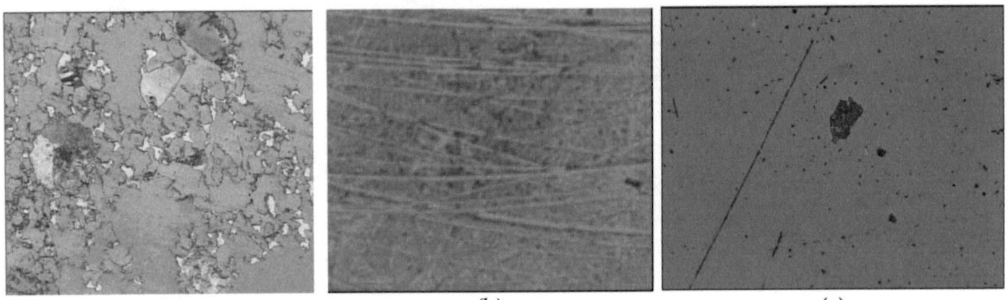

Figure 17. Attacked electrochemical corrosion surface bilayer micrographs a) 40% CH4, b) 60% CH4 and c) 80% CH4.

The inhibition of the cathodic and anodic processes can be obtained from the data in table 2.

Substrate	Ecorr (mV)	Icorr (A)	Corrosion Rate (mpy)
AISI 420 steel	-407,0	10x10-6	4,587
40% CH4	-440,0	29x10-6	13,25
60% CH4	-293,0	127x10-9	0.05833
80% CH4	-165,0	281x10-9	0.1286

Table 2. Data from the electrochemical corrosion test on probes without and with coatings.

The corrosion speed expressed as corrosion current Icur. was calculated by using the extrapolation Taffel method. The 40% CH4, according data in table 2 shows a corrosion potential more negative as compared with the other two, indicating a system more active and thus lesser corrosion resistance.

The roughness measured on the plain steel was 0.02 μm, whilst that of the W/WC bilayers was 0.24 μm, 0.26 μm and 0.29 μm for 40% CH4, 60% CH4 and 80% CH4, respectively.

9. Tribological properties

The films were undergone to the wear test using the pin on disk technique. Figure 18 shows the curves friction coefficient versus sliding distance. These curves correspond to the coatings W/WC grown with 40, 60 and 80 % of CH4, respectively. As it is evident they exhibit three important and well defined regions. The first zone between 0 m and 500 m, of the displacement when a sudden increase in the friction coefficient is present, due to the initial contact between the pin and the probe, the intermediate region between 500 m and 1100 m of the displacement a diminution is present due to a more flat surface, and finally in region three when the coefficient is stable.

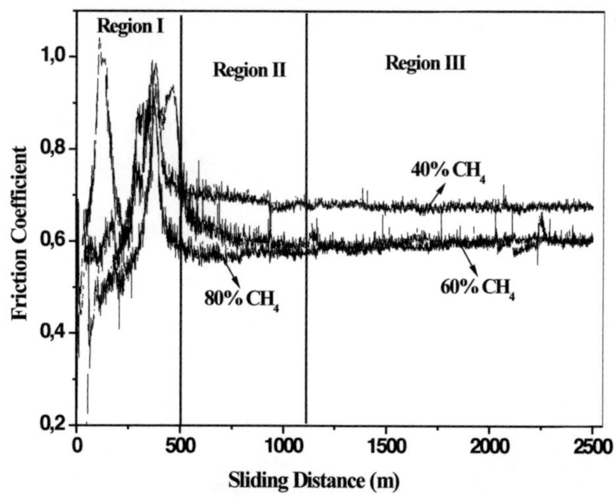

Figure 18. Friction coefficient of hard coatings W/WC grown with different CH4 on to steel AISI 420 as a function of sliding distance.

The graph reveals a slight increase in the friction coefficient that is associated with a process of micro- welding that are formed and destroyed as the pin slips on the probe and as the material wear out allowing the formation of new surfaces. These new surfaces have a tribological behavior different from that of the original surface.

The test showed that the behavior of the AISI 420 steel without a coating and the one belonging to the film grown with 40% of CH4, was similar, the third region show a stabilized friction coefficient of 0.679. The curve corresponding to the bilayer behavior W/WC grown with 40% CH4, shows in the second and third region a higher friction coefficient value as compared with the films grown with 60 y 80% CH4.

Figure 18 shows the curve corresponding to the behavior of the bilayer W/WC at 60% CH4. The friction coefficient stabilizes at 0.602, while the curve corresponding to the bilayer grown with 80% CH4 showed a similar behavior to which of the bilayer with 60% CH4, but its friction coefficient stabilized at 0.585.

The test pin-on-disk traces were analyzed by using an image analyzer. In figure 19 the observed trace on the AISI 420 steel is almost homogeneous, but wear is too severe.

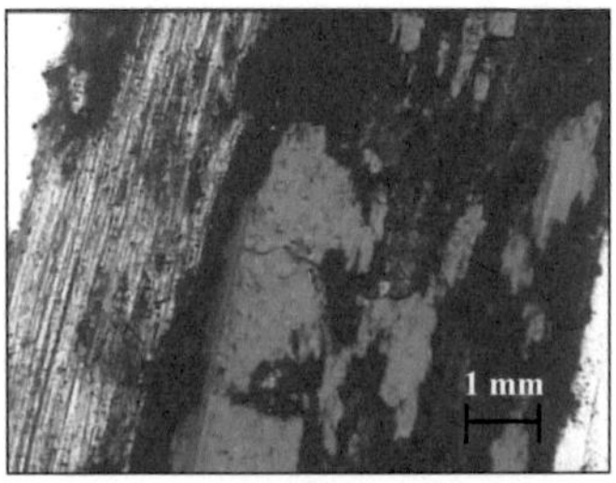

Figure 19. Photomicrograph showing the wear patterns AISI 420 steel.

The micrographs of figure 20 exhibit the wear surfaces of the bilayers. Figure 20a corresponds to W/WC 40% CH4 this bilayer shows an excessive wear, in some places the peeling of the film was observed.

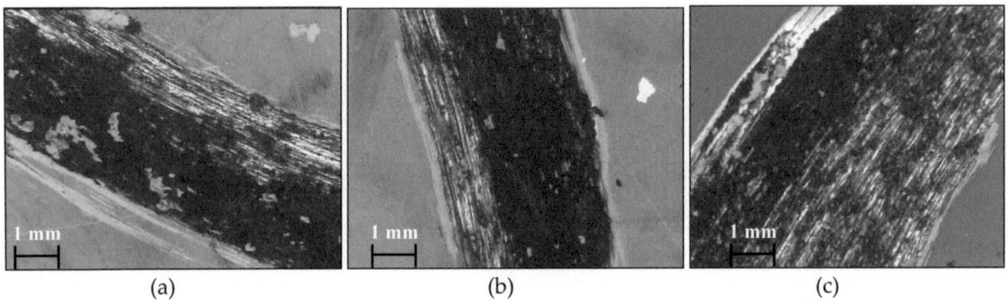

Figure 20. Trace micrographs on AISI 420 coated steel with W/WC. a) 40% CH4, b) 60% CH4 and c) 80% CH4 bilayers

Figure 20b and 20c corresponding to 60% and 80% of CH4 respectively, can be observed a homogeneous wear without the peeling of the films.

The above results showed that the bilayers grown with 80% CH4 present lower friction coefficient and higher wear resistant, when compared with the bilayers grown with 40% and 60% CH4, This allows concluding that as the methane percentage increases in the gas mixture a better film adhesion to substrate is presented. Therefore it's sure that the structure was principally composed by the cubic WC1-x phase with a preferred (111) orientation imbibed in an amorphous carbon matrix.

Sample	Initial Weight (mg)	Loss Weight (mg)	Friction Coefficient
AISI 420 Steel	4.3345	1.5	0.679
40% CH4	5.2351	0.98	0.679
60% CH4	5.1812	0.12	0.602
80% CH4	4.9566	0.10	0.585

Table 3. Data about the material lost and friction coefficient during the pin on disk test.

The wear of the bilayers is summarized in table 3. The higher value corresponds to that one grown at 40% CH4, while the one of those with 60% and 80% CH4, have lower values.

10. Conclusions

TiC monolayers and Ti/TiN/TIC/TiCN multilayers were grown through the pulsed-laser technique on AISI 4340 steel. It was determined that the adhesion of the films is quite acceptable. When the tension test was applied there was no detachment of the substrate on the monolayers or the multilayers. For AFM, it was found that the multilayers presented finer grain size, as well as lower roughness than the monolayers. The wear tests revealed that the multilayers showed better resistance than the TiC monolayers. In corrosion tests, the multilayers presented a lower mass loss and the film was not attacked by corrosion in its totality, keeping in mind that the medium was quite aggressive. This indicates that in engineering applications, where corrosive environments are present in the workplace, the recommendation is to use multilayered coatings on AISI 4340 steel, or any other with similar chemical properties.

In both optical microscopy images and SEM images we observed that samples treated with the GMAW welding process were most affected by intergranular corrosion, in segregated zones within the welt; whereas in samples welded with SMAW and GTAW processes the corrosion advance was not as significant. Perhaps the cause for the corrosion phenomenon described is due to segregation levels present in the solder welts, which generated – according to the results – greater presence of carbon in the segregated zones and, additionally, a galvanic pair is registered between the segregated and non-segregated zones; thus, accelerating the corrosion phenomenon.

The W/WC bilayers were grown on AISI 420 stainless steel by using the magnetron reactive sputtering technique. The corrosion tests have shown that the steels coated with 60% and

80% CH4 in the gas mixture exhibit high corrosion resistance. The wear tests established the steel coated with W/WC bilayers with 60% and 80% CH4, in the gas mixture were more wear resistance than the others. These results will be validate on the next time at industrial scale through the coating on cutting tools at the pilot magnetron sputtering facilities of CDT ASTIN and their corresponding cutting tests.

Acknowlegment

This work was financed by the office of Research and Technological Development at the *Universidad Autónoma de Occidente,* Cali Colombia, and supported by COLCIENCIAS under the program Excellence Center for Novel Materials, contract No. 0043-2005. The authors wish to thank the Tribologic and Surfaces Laboratory at the *Universidad Nacional,* Medellín Colombia, directed by Dr. Alejandro Toro.

The authors thank Ingenio Providencia – the sugar mill – for facilitating the field samples at their facilities. Characterization of samples was done in the Materials lab at UAO and at the Plasma lab of Universidad Nacional de Colombia, Manizales.

This work was supported by COLCIENCIAS under the program Excellence Center for Novel Materials, CENM contract No. 0043-2005. The growth of the bilayers was carried out at pilot facilities of SENA-ASTIN and the films characterization at the Marco Fidel Suarez School.

Author details

N. A. de Sánchez[1,2,5*], H. E. Jaramillo[1,2,5], H. Riascos[3,5], G. Terán[1], C. Tovar[1,2], G. Bejarano G.[5,6,7], B. E. Villamil[1], J. Portocarrero[1,3], G. Zambrano[4,5] and P. Prieto[4,5]

*Address all correspondence to: nalba@uao.edu.co

1 Science and Engineering of Materials Group, Colombia

2 Mechanical Engineering Program, Universidad Autónoma de Occidente, Cali, Colombia

3 Department of Physics, Universidad Tecnológica de Pereira, Pereira, Colombia

4 Thin Film Group, Department of Physics, Universidad del Valle, Cali, Colombia

5 Excellence Center for Novel Materials, Universidad del Valle, Cali, Colombia

6 Group for Engineering and Materials Development, CDT ASTIN-SENA, Colombia

7 Cali-Colombia, Centre for Research, Innovation and Development of Materials CIDEMAT, Universidad de Antioquia, Medellín,, Colombia

References

[1] A. Y. Liu and M. Cohen, Physics Rev. B 41, 10727 (1990)

[2] D. Sanders, A. Anders, Surf. Coat. Technol, 133 (2000) 78

[3] S. Veprek, J. Vac. Sci. Technolo. A 17, (199) 2401

[4] Y. Kusano, J.E. Evetts, R.E. Somekh, and I.M. Hutchings, Thin Solid Films Elsevier 56, (1998), 1272

[5] Y. Kusano, Z. H. Barber, J. E. Evetts and I. M. Hutchings, Surface and Coatings Technology 124 (2000) 104

[6] M. Show Wong, A. Lefkow, and W. Sproul, Group Program Prospectus BIRL Northwestern University Illinois 60201 (1999)

[7] M. Teter, MRS Bulletin 22, January (1998)

[8] A. Toro, A. Sinatora, D.K. Tanaka, A.P. Tschiptschin, Wear 251 (2001) 1257

[9] Diana López, Carlos Sánchez, Alejandro Toro, Wear 258 (2005) 684

[10] U. Helmersson, S. Todorava, S. A. Barnett, J. Sundgren, J. Appl. Phys. 62 (1887) 481

[11] P. B. Mirkarimi, L. Hultman, S.A. Barnett, Appl. Phys. Lett. 57 (1990) 2654

[12] H. Ljungcrantz, C. Engström, L. Hultman, M. Olsson, X. Chu, M. S. Wong, W. D. Spoul, J. Vac. Sci Technol. A16 (1998) 3104

[13] K. K. Shih, D. B. Dove, Appl. Phys. Lett. 61 (1992) 654

[14] J.P. Tu, L.P. Zhu, H.X. Zhao, Surf. Coat. Technol., 122 (1999) 176

[15] Wilson, S. and Alpas, A.T., Wear, 245, (2000) 229

[16] X. Wang. A. Kolitsch, W. Möller, Appl. Phys Lett. 71 (1997) 1951

[17] K. L. Mittal, Adhesion measurement of films and coatings, Utrecht (1995)

[18] H.M. Clark, K.K. Wong, Wear 186 (1995) 454

[19] Bhowmick S., Kale A. N., Jayaram V., Biswas S. K., Thin Solid Films, 436 (2003) 250.

[20] Smith, William F. Fundamentals of Science and Engineering of Materials. 3rd edition. McGraw Hill, Interamericana. pp. 595 – 636; 1998.

[21] http://www.corrosiondoctors.org/principles/theory.htm. Último acceso 20 de Mayo de 2004.

[22] Botia F., J. S. Ingeniería de Corrosión. Instituto Nacional del Acero. Bogotá, D. E. 12/1985.

[23] http://ionis.com.ar/agua/corrosion.htm Último acceso 27 de Marzo de 2004.

[24] Roberge, Pierre R. Handbook of Corrosion Engineering. Editorial McGraw Hill; 2000.

[25] Reina Gómez, Manuel. Soldadura de los Aceros Aplicaciones, Second Edition. Graficas Lormo; 1988.

[26] Smith, William F. Fundamentals of Science and Engineering of Materials. 3rd edition. McGraw Hill, Interamericana; 2000.

[27] Document by Ingenio Providencia Titled: Sulfitación del jugo; 2001.

[28] Kuo, Sindo. Welding Metallurgy, Second Edition. Wiley-Interscience. A John Wiley & Sons, Inc., Publication. 2003

[29] Otero Huerta, Enrique. Corrosión y Degradación de Materiales; Editorial Síntesis S.A.; 1997.

[30] Easterling, Kenneth. Introduction to the Physical Metallurgy of Welding. First Published. Editorial Dutterworths; 1983.

[31] Okazaki. Y. Ito, K. Kyo, T. Tateishi, Mat. Sci. Eng. A213 (1996) 138.

[32] J.A. Sue, T.P. Chang, Surf. Coat. Technol. 76-77 (1995) 61.

[33] L.J. Yang, Determination of the wear coefficient of tungsten carbide by a turning operation, Wear 250 (2001) 366–375.

[34] J.A. Sue, T.P. Chang, Surf. Coat. Technol. 76-77 (1995) 61.

[35] H. Dimigen, H. Hübsch, Carbon containing sliding layer, USPatent 4 525 417 (1985).

[36] H. Dimigen, H. Hübsch, R. Memming, Appl. Phys. Lett. 50 (1987) 1056.

[37] E.A. Almond, M.G. Gee, Results from a U.K. inter-laboratory Project on dry sliding wear, Wear 120 (1987) 101–116.

[38] Y. Liu, M. Gubisch, W. Hild, M. Scherge, L. Spiess, Ch. Knedlik, J.A. Schaefer, Nanoscale multilayer WC/C coatings developed for nanopositioning part II: friction and wear, Thin Solid Films 488 (2005) 40-148.

[39] C. Rincón, G. Zambrano, P. Prieto, Propiedades mecánicas y tribológicas de monocapas de W-C, DLC y multicapas de WC/DLC, Revista Colombiana de Física, vol. 35, No. 1 (2003).

[40] A. Voevodin, J.P. O`Neill, J.S. Zabinski, tribological performance and tribochemistry of nanocrystalline WC/ DLC composites, Thin Solid Films vol. 342, Issue 1-2 (1999) 194-200.

Section 2

Corrosion Science and Technology

Chapter 4

Corrosion Behaviour of Cold-Deformed Austenitic Alloys

Wojciech Ozgowicz, Agnieszka Kurc-Lisiecka and Adam Grajcar

Additional information is available at the end of the chapter

http://dx.doi.org/10.5772/53590

1. Introduction

Austenitic stainless steels are the most important group of corrosion-resistant metallic materials finding widespread industrial application. The material has been employed in many applications ranging from pharmaceutical equipment to piping in the nuclear reactors. The microstructure of the austenitic stainless steels is composed of the monophase austenite (γ), and the steels generally have low (~16 mJm^{-2}) or medium (~20 mJm^{-2}) values of the stacking fault energy (SFE). Plastic deformation of these steels leads to a phase transformation from paramagnetic austenite into ferromagnetic martensite. Depending on the chemical composition, stacking fault energy, phase stability and deformation conditions (temperature, strain rate and strain value) different transformations can take place (i.e., $\gamma \rightarrow \varepsilon$, $\gamma \rightarrow \varepsilon \rightarrow \alpha'$ or $\gamma \rightarrow \alpha'$) in metastable austenitic steels. At the early stage of deformation, shear bands consisting of stacking faults and deformation twins are formed, promoted by the low SFE of these steels. The ε-martensite phase is formed by overlapping stacking faults, and therefore, it is finely dispersed and its structure is heavily faulted. The crystal structure of α'-martensite is body-centered cubic, and it is usually nucleated at the intersections of shear bands. The volume fractions and distribution of individual phases influence mechanical properties and corrosion resistance of these steels.

Literature data show (Zhilin et al., 2006; Kumar et al., 2007; Rutkowska-Gorczyca et al., 2009) that the cold deformation of Cr-Ni steels worsens their corrosion resistance. It can be explained for several ways. The delivering to material additional energy as a result of the induction of external tensions is generally accepted as one of the causes of the lowering of corrosion resistance. It leads in consequences to decreasing the thermodynamic durability of the material. The second reason is connected with the crystallographic structure of cold-deformed material which shows the characteristic orientation of grains (texture). Crystallographic texture plays an important role altering pitting mechanism, possibly by reducing favorable sites for corrosion attack and formation of Cr-rich passive film favored by high-density close-packed planes oriented parallel to the rolling surface (Kumar et al.; 2005).

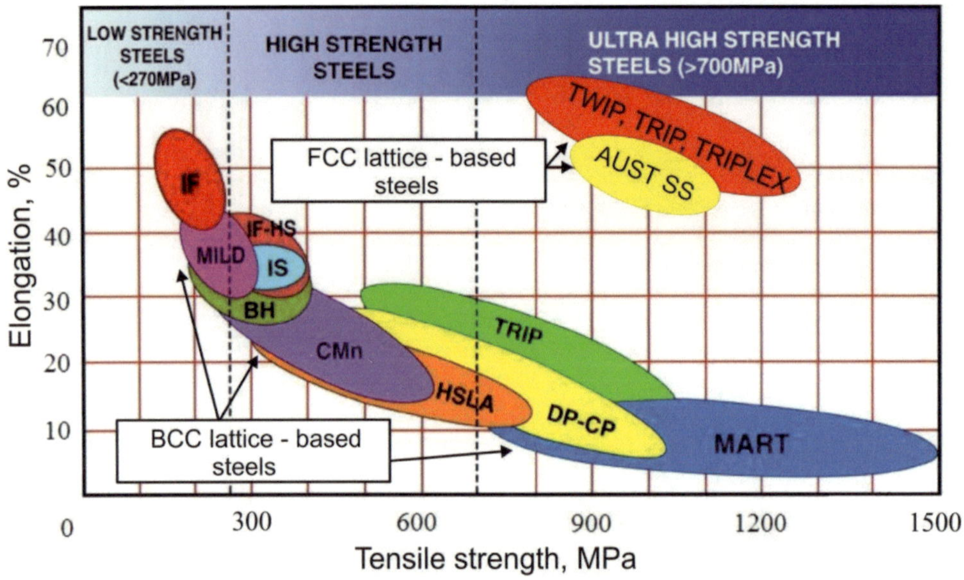

Figure 1. Schematic comparison of mechanical properties of conventional and advanced BCC lattice-based steels with Cr-Ni (AUST SS) and high-Mn (TWIP, TRIP, TRIPLEX) austenitic alloys (International Iron & Steel Institute, 2006).

In some applications the Cr-Ni alloys would be very useful due to their excellent combination of strength, ductility and corrosion resistance but their application is limited because of the high cost. For example, it relates to their potential application in the automotive industry or for cryogenic applications. Austenitic microstructure can be also formed by manganese alloying at much reduced cost. The manganese content guaranteeing an uniform austenitic microstructure is equal to about 25% for carbon contents between 0.04 and 0.1% (Frommeyer et al., 2003; Graessel et al., 2000). The manganese content can be decreased to about 17% for steels with higher carbon concentrations (up to 0.8 wt.%) (De Cooman et al., 2011; Ghayad et al., 2006; Jimenez & Frommeyer, 2010). Sometimes, high-manganese steels contain up to 4% Al and/or Si (Frommeyer et al., 2003; Graessel et al., 2000), chromium (Hamada, 2007; Mujica Roncery et al., 2010) or microadditions of Nb, Ti and B (Grajcar et al., 2009, 2010a, 2010b; Huang et al., 2006). Independently of a chemical composition, different grades of both high-manganese and Cr-Ni austenitic steels offer an exceptional combination of high-strength and ductility compared to conventional and advanced BCC lattice-based steels (Fig. 1). The source of unique mechanical properties and technological formability of austenitic alloys is the great susceptibility of γ phase on plastic deformation, during which dislocation glide, mechanical twinning (TWIP – Twinning Induced Plasticity) and/or strain-induced martensitic transformation (TRIP – TRansformation Induced Plasticity) can occur.

The key to obtain the mechanical properties range in Fig. 1 is the high work hardening rate characterizing the plastic deformation of austenitic alloys. The high level of ductility is a result of delaying necking during straining. In case of the local presence of necking, strain-induced martensitic transformation occurs in such places (in TRIP steels) or deformation twins are preferably generated in locally deformed areas (in TWIP steels). It leads to intensive local strain hardening of the steel and further plastic strain proceeds in less strain-hardened adjacent zones. The situation is repeated in successive regions of the material what finally leads to delaying necking in a macro scale and high uniform and total elongation (De Cooman et al., 2011; Frommeyer et al., 2003; Grajcar, 2012). The shear band formation accompanied by dislocation glide occurs in deformed areas of TRIPLEX steels and the SIP (Shear band Induced Plasticity) effect is sustained by the uniform arrangement of nano size κ-carbides coherent to the austenitic matrix (Frommeyer & Bruex, 2006).

2. Corrosion behaviour of austenitic alloys

2.1. Cr-Ni austenitic steels

Corrosion resistance of stainless steels is achieved by dissolving a sufficient content of chromium in iron to produce a coherent, adherent, insulating and regenerating chromium oxide protective film (Cr_2O_3) on a surface. The stainless character occurs when the concentration of Cr exceeds about 12 wt%. The passive film of chromium oxide formed in air at room temperature is only about 1-2 nm. However, this is not adequate to resist corrosion in acids such as HCl or H_2SO_4 (Azambuja et al., 2003; Kurc et al., 2010). In environments containing chloride the austenitic steels are susceptible to localized corrosive attacks, such as pitting corrosion, intergranular corrosion and stress corrosion cracking (SCC) (Osawa & Hasegawa, 1981; Ningshen et al., 2010).

Pitting corrosion is the result of the local destruction of the passive film and subsequent corrosion of the steel below. It generally occurs in chloride, halide or bromide solutions. It can be initiated at a fault in the passive layer or at a surface defect. Pitting is considered to be autocatalytic in nature; once a pit starts to grow, the conditions developed are such that further pit growth is promoted. The anodic and cathodic electrochemical reactions that comprise corrosion separate spatially during pitting. The local pit environment becomes depleted in cathodic reactant (e.g., oxygen), which shifts most of the cathodic reaction to the boldly exposed surface where this reactant is more plentiful. The pit environment becomes enriched in metal cations and anionic species such as chloride, which migrate into the pit to maintain charge neutrality by balancing the charge associated with the cation concentration (Frankel, 1998). Even in a neutral solution, this can cause the *pH* to drop locally to 2 or 3, thereby preventing the regeneration of the passive layer. In the passive condition, the current density is of the order of $\mu A\ cm^{-2}$. However, it may exceed $1A\ cm^{-2}$ in the pit. The reason why the current density is so large in the pit is that the anodic region is a very small area when compared with the cathodic part (the steel free of the pits). For a given corrosion current, this greatly exaggerates the corrosion rate at the pits. Similarly, the concentration of chloride ions in the vicinity of a pit can be thousands of times greater than that in the solution as a whole (Otero et al., 1995; Padro et al., 2007).

The anodic dissolution of steel leads to introduction of positive metal ions (M^+) into solution, which causes migration of Cl^- ions. In turn, metal chloride reacts with water according to the reaction (1):

$$M^+Cl^- + H_2O \rightarrow MOH + H^+Cl^- \tag{1}$$

This causes the *pH* to decrease. The cathodic reaction, on the surface near the pit follows (2):

$$O_2 + 2H_2O \rightarrow MOH + 4OH^- \tag{2}$$

Pitting is mainly associated with microscopic heterogeneities at a surface rather than macroscopic physical features of a component. Wet and humid environments containing chloride ions can cause pitting corrosion and crevice corrosion of austenitic stainless steel components. Chloride ions are known to be potent corrosion enhancers and localized adsorption of chloride ions can act as prenuclei for pitting. Pits can also nucleate at carbides, grain boundaries and other material inhomogenities on the metal surface. The presence of moisture in the environment can also facilitate the electrolytic path for the chloride ions (Khatak & Raj, 2002).

The corrosion behaviour of hydrogen-containing austenitic stainless steels has also been studied and it has become clear that the intergranular corrosion (Sunada et al., 2006) and the general corrosion (Osawa & Hasegawa, 1981; Sunada et al., 2006) were accelerated by hydrogenation. This phenomenon was noticed by the authors as "anomalous corrosion". Anomalous general corrosion of austenitic stainless steels is observed when steels contain stress-induced martensite. These BCC regions become anodically active sites due to hydrogenation and are attacked selectively. On the other hand, the stress induced martensite plays an important role during crack initiation and propagation processes when stress corrosion cracking in H_2SO_4-NaCl solution occurs. Osawa & Hasegawa (1981) investigated the corrosion behaviour of hydrogen-containing deformed austenitic stainless steels and observed that hydrogenation increased the corrosion rate in 5N-H_2SO_4 solution when the steel contained stress-induced martensite. They found that the corrosion rate increases with increasing the volume fraction of martensite and hydrogenation facilitates this tendency. The similar effect of stress induced martensite on stress corrosion cracking of the 304 and 310 stainless steels was observed by Qiao & Luo (1998).

2.2. High-Mn austenitic steels

The research on high-Mn-Al alloys for cryogenic applications that were supposed to substitute Cr-Ni steels was carried out in the eighties of the last century (Altstetter et al., 1986). The role of manganese is to replace Ni and to obtain an austenitic microstructure, whereas aluminium has a similar impact as chromium. Improvement of corrosion resistance by Al consists in formation of thin stable layer of oxides. Alstetter et al. (1986) found that high-Mn-Al alloys show inferior corrosion resistance than Cr-Ni steels and they can be used as a substitute only in some applications. Nowadays, the application of high-Mn austenitic steels in the automotive industry rises and some works are undertaken to improve the corrosion resistance of these alloys.

The addition of 25% Mn to mild steels was found to be very detrimental to the corrosion resistance in aqueous solutions (Zhang & Zhu, 1999). The Fe-25Mn alloy was difficult to passivate, even in 1M Na_2SO_4 solution. With increasing Al content up to 5% of the Fe-25Mn-Al steel, the anodic polarization curves exhibit a stable passivation region in Na_2SO_4 solution, but it shows no passivation in 3.5 wt% NaCl solution. Recently, corrosion resistance of Fe-0.05C-29Mn-3.1Al-1.4Si steel in acidic (0.1M H_2SO_4) and chloride-containing (3.5 wt% NaCl) environments was investigated by Kannan et al. (2008). They found that the high-Mn-Al-Si steel has lower corrosion resistance than an ultra deep drawing ferritic steel, both in acidic and chloride media. The corrosion resistance of the high-manganese steel in chloride solutions is higher compared to that observed in acidic medium.

The behaviour of Fe-0.2C-25Mn-(1-8)Al steels with increased concentration of Al up to 8% in 3.5 wt% NaCl was also investigated by Hamada (2007). It was reported that the corrosion resistance of tested steels in chloride environments is pretty low. The predominating corrosion type is the general corrosion, but locally corrosion pits were observed. In steels including up to 6% Al with homogeneous austenite structure, places where the pits occur are casually, whereas in case of two-phase structure, including ferrite and austenite (Fe-0.2C-25Mn-8Al), they preferentially occur in α phase. It was found that complex addition of Al and Cr to Fe-0.26C-30Mn-4Al-4Cr and Fe-0.25C-30Mn-8Al-6Cr alloys increases considerably the general corrosion resistance, especially after anodic passivation ageing of surface layers in an oxidizing electrolytic solution (Hamada, 2007). Cr-bearing steels passivated by nucleation and growth of the passive oxide films on the steel surface, where the enrichment of Al and Cr and depletion of Fe and Mn have occurred. The positive role of Cr in obtaining passivation layers in 0.5M H_2SO_4 acidic solution was recently confirmed by Mujica Roncery et al. (2010) in Fe-25Mn-12Cr-0.3C-0.4N alloy.

2.3. Effect of cold deformation

Stainless steels are usually subjected to different levels of cold working during final manufacturing stages. Cold deformation affects the corrosion resistance of stainless steels because planar dislocation arrays (Oh & Hong, 2000) and deformation twins (Lee et al., 2007) are introduced. Barbucci et al. (2002) reported that the passive currents in both sulfate + chloride and sulfuric acid solutions significantly increased with increasing the degree of cold working of the 304 type stainless steel. The pitting susceptibility also increased with cold working, especially evident as the chloride concentration increased. The dependence of the pitting potential on cold working was explained using the bilayer model, the sulfate ingresses in the passive film during anodic oxidation forming a coulombic barrier against chloride penetration. On the basis of microscopic examination after long time polarization in 1M H_2SO_4 acidic solution it was found, that the surface profile of the passive film depended on the metallurgical structure of a metallic substrate. The higher passive currents and increased susceptibility to pitting corrosion of the work hardened samples were explained by the formation of much more defective oxides during its anodic oxidation, with easy paths that enhanced sulfate ingress. The growth of such oxides was related to the formation of defects in the grains and more defective interfaces in the bulk material, resulting from the accumulation of internal stresses during cold rolling (Barbucci et al., 2002).

Efforts have been also made to clarify the relationship between cold working and a sensitization process. Briant (1982) observed a transgranular attack in the 304 type stainless steel due to strain-induced martensite. However, a similar corrosion attack has been also noted in the deformed 316 type stainless steel consisting of an uniform austenitic microstructure. The cold working effect in the 316 type steel has been attributed to higher diffusivity of chromium and the lower free energy barrier to carbide nucleation at grain boundaries in the deformed microstructure. The acceleration of sensitization due to cold deformation could be also related to the effects of point defects and microstructural sinks on diffusion (Fu et al., 2009).

The influence of cold working, in particular the amount of α' martensite on the behaviour of AISI 321 stainless steel in 3.5 wt% NaCl was studied by Xu et al. (2004). They showed that when the content of martensite was less than 6% and more than 22%, the pitting sensitivity increased. However, when the martensite content was between 6 and 22%, the sensitivity decreased with increasing its content. Fang et al. (1997) reported that the corrosion potential of the martensite phase was more negative than that of austenite and this is the primary reason for the selective corrosion of martensite. They also demonstrated that the pitting potential of the austenite phase was more noble than that of the martensite phase and the stable passive current of austenite was lower compared to martensite.

Sunada et al. (1991) found that the number of pits formed on AISI 304 austenitic stainless steel in H_2SO_4-NaCl solution had a direct relationship with the martensite content. The effect of α' phase on the number of pits depended upon the NaCl concentration, temperature and potential. In the case of high NaCl concentration, the degree of pitting increased linearly with increasing volume fraction of martensite. Under high temperature and high anodic potential conditions, the number of pits was almost constant in the range of volume fractions greater than 50%. The corrosion rate of the martensitic phase was about 1.65 times that the magnitude of the austenitic phase at volume fractions of martensite below 50%. At volume fractions higher than 50%, the corrosion rate was more enhanced. Peguet et al. (2007) studied the influence of cold rolling with 10, 20, 30 and 70% reduction on the pitting corrosion of AISI 304 stainless steels. Electrochemical tests were carried out in 0.5M NaCl. They showed, that the pit propagation rate increases monotonously with cold rolling and the pit repassivation ability decreases (leading to a larger number of stable pits), suggesting that the overall dislocation density is the most important controlling factor.

The influence of cold deformation on corrosion behaviour in 3.5 wt% NaCl in high-Mn steels was studied by Ghayad et al. (2006). They found that the 0.5C-29Mn-3.5Al-0.5Si steel shows no tendency to passivation, independently on the steel structure after heat treatment (supersaturated, aged or strain-aged). Higher corrosion rate of cold worked specimens compared to that in the supersaturated state, was a result of faster steel dissolution caused by annealing twins, which show a different potential than the matrix. The highest corrosion rate was observed in strain-aged samples, as a result of ferrite formation, which creates a corrosive galvanic cell with the austenitic matrix. High-Mn steels containing hydrogen-induced or strain-induced ε/α' martensite are also susceptible to hydrogen embrittlement (Lovicu et al., 2010) and delayed fracture (De Cooman et al., 2011; Shin et al., 2010).

3. Experimental procedure

3.1. Material

The chapter addresses the corrosion behaviour of cold-deformed Cr-Ni and high-Mn austenitic steels in chloride and acidic media. The chemical composition of the investigated steels is presented in Table 1.

Steel grade	C	Mn	Si	P	S	N	Cr	Ni	Mo	Al	Nb	Ti
X5CrNi18-8	0.030	1.31	0.39	0.030	0.004	0.0440	18.07	8.00	0.25	-	-	-
X6MnSiAl26-3-3	0.065	26.0	3.08	0.004	0.013	0.0028	-	-	-	2.87	0.034	0.009

Table 1. Chemical composition of the investigated steels, wt.%

The microstructure of the stainless steel in a delivery state (after solution heat treatment) is shown in Fig. 2. The X5CrNi18-8 steel exhibits a homogeneous austenite structure with annealing twins (Fig. 2a). The X-ray diffraction analysis confirms the single-phase structure of the steel (Fig. 2b). The high-Mn steel is characterized also be the single-phase austenitic microstructure containing many annealing twins (Fig. 3). The grain sizes of both steels are similar, i.e., from 15 to 30 μm. The grains of X6MnSiAl26-3-3 steel are slightly elongated as a result of hot rolling followed by solution heat treatment from a temperature of 850°C.

The stainless steel used for the investigation was delivered in a form of sheet segments with dimensions 700x40x2 mm, whereas the high-Mn steel as 340x225x3.2 mm sheet specimens. The Cr-Ni steel specimens were subjected to cold rolling with the various amount of deformation ranging from 10%, 20%, 30%, 40%, 50% to 70%. The cold rolling was conducted at room temperature keeping a constant direction and a side of the rolled strip.

Cold deformation of high-manganese steel specimens was applied by bending at room temperature. The samples with a size of 10x15 mm and a thickness of 3.2 mm were bent to an angle of 90° with a bending radius of 3 mm.

The immersion tests and potentiodynamic polarization tests were used to assess the corrosion resistance of austenitic steels in the delivery state and after cold deformation.

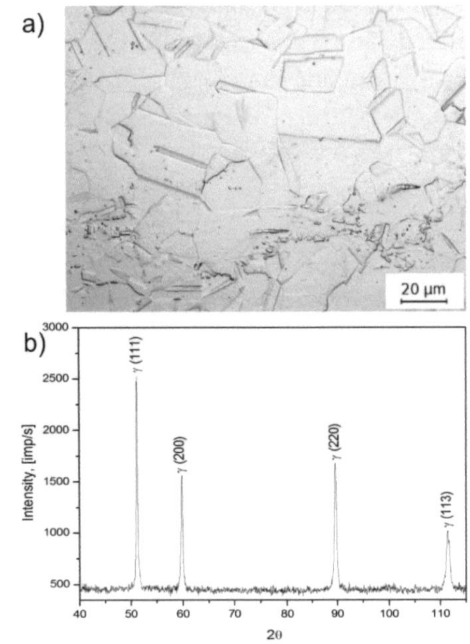

Figure 2. Austenitic microstructure with annealing twins of X5CrNi18-8 steel in a delivery state (a) and X-ray diffraction pattern (b).

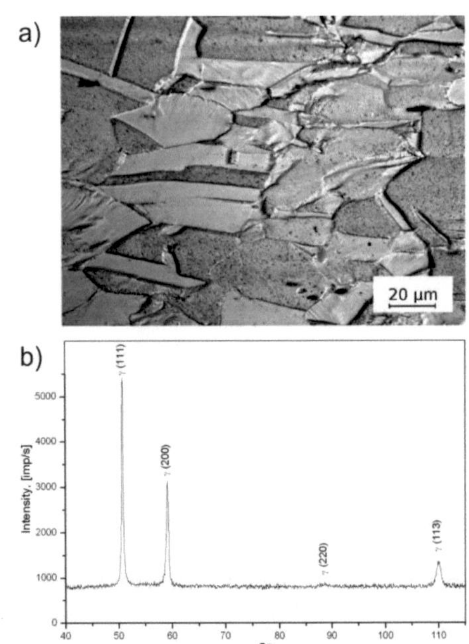

Figure 3. Austenitic microstructure with annealing twins of X6MnSiAl26-3-3 steel after the thermo-mechanical rolling and immersion in 1N H_2SO_4 (a) and X-ray diffraction pattern (b).

3.2. Immersion tests

The corrosion resistance of the investigated steels was assessed using immersion tests in sulfuric acid and chloride solutions. 3.5 wt% NaCl solution was used for both steels whereas 3.5N H_2SO_4 and 1N H_2SO_4 solutions were used respectively for Cr-Ni and high-Mn steels for the sake of high difference in a corrosion progress. Corrosion tests of the austenitic stainless steel were carried out on samples with dimensions 20×15 mm, whereas 10x15 mm samples were cut for the high-Mn steel. The specimens were taken from the material in the delivery state as well as from sheets after all subsequent deformation stages. Before starting the analysis the samples were washed in distilled water, ultrasonically cleaned in acetone and finally cleaned in 95.6% ethanol. The specimens were weighed with the accuracy of 0.001g and put into solution at room temperature. The time of the test for the stainless steel was equal to 87 days for 3.5 wt% NaCl solution and 25 days in the case of 3.5N H_2SO_4. The majority of samples was dipped entirely whereas a few samples were immersed only partially to compare dipped and original surfaces. Taking into account much faster corrosion rate of the high-Mn steel, it was put into 3.5 wt% NaCl and 1N H_2SO_4 solutions for 4 days. After the tests the specimens were weighed and analyzed using optical microscopy and SEM. Corrosion loss was calculated in a simple way as the difference between initial and final mass of the samples. Percentage mass decrement was also calculated.

Metallographic observations of non-metallic inclusions and corrosion pits were carried out on polished sections, whereas the microstructure observations on specimens etched in nital (high-Mn steel) or in chloroazotic acid (Cr-Ni steel). The investigations were performed using LEICA MEF 4A optical microscope, with magnifications from 100 to 1000x. Fractographic investigations were carried out using scanning electron microscope SUPRA 25 (Zeiss) at the accelerating voltage of 20kV. In order to remove corrosion products, the specimens were ultrasonically cleaned before the analysis.

3.3. Potentiodynamic polarization tests

Investigation of the electrochemical corrosion behaviour of the stainless steel samples was done in a PGP 201 potentiostat using a conventional three-electrode cell consisting of a saturated calomel reference electrode (SCE), a platinum counter electrode and the studied specimens as the working electrode (Fig. 4). The results of the electrochemical corrosion behaviour of the high-Mn steel are presented elsewhere (Grajcar, 2012). The tests were carried out at room temperature in the electrolyte simulating sea water (3.5 wt% NaCl) on specimens of size 30×20mm with an exposed sample area of about 1cm^2. Registering of anodic polarization curves was conducted at the rate equal to 1mV/s. The measurement of corrosion potential was realized in time of 60 min. According to registered curves the corrosion potential (E_{cor}), polarization resistance (R_p) and corrosion current density (i_{cor}) were determined. In order to calculate the corrosion current, the Stern-Geary equation was used (Baszkiewicz & Kamiński, 1997):

$$R_p = \frac{b_k \cdot b_a}{2.3 \cdot i_{cor}(b_a + b_k)} \left[k\Omega \cdot cm^2 \right] \quad (3)$$

where: b_k - the slope coefficient of the cathodic Tafel line, b_a - the slope coefficient of the anodic Tafel line, i_{cor} - the corrosion current density [μA/cm^2], R_p - the polarization resistance [kΩ cm^2].

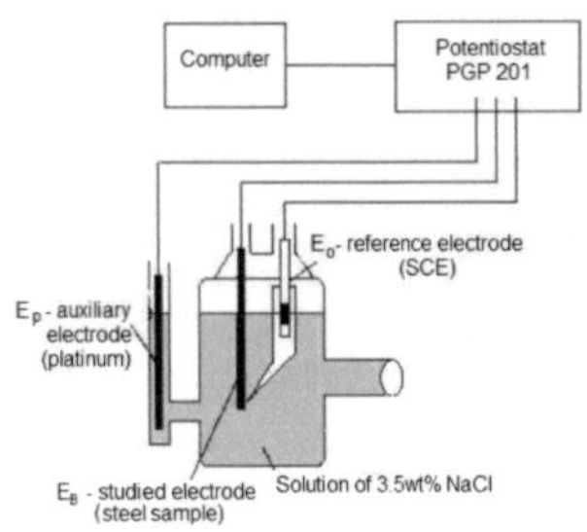

Figure 4. Schematic showing of the corrosion resistance investigation set.

4. Results and discussion

4.1. Cr-Ni austenitic steel

4.1.1. Results of immersion tests

Results of the immersion tests in two media for non-deformed samples are given in Table 2. After immersion in 3.5N H_2SO_4 the X5CrNi18-8 steel showed a percentage mass decrement about 38%. The mass loss of specimens dipped in 3.5 wt% NaCl is over 100 times lower despite longer time of the immersion compared to specimens dipped in the acidic solution. Štefec & Franz (1978) observed the similar order of magnitude of corrosion progress for the AISI 304 steel. The difference is due to different corrosion mechanisms. When the solution is acidic, the corrosion process is running according to hydrogen depolarization, whereas in chloride media the specimens are corroding with oxygen depolarization.

X5CrNi18-8 steel	**Corrosion medium**	
State	3.5N H_2SO_4	3.5 wt% NaCl
non-deformed	38.2 ± 6.3	0.24 ± 0.08
50% cold deformed	67.5 ± 9.8	0.72 ± 0.23

Table 2. Mean percentage mass loss of samples in an initial state and 50% cold-deformed after the immersion tests, %

Many micropores and corrosion pits along the whole specimen surface were observed in the X5CrNi18-8 steel immersed in 3.5N H_2SO_4 (Fig. 5). Slightly smaller corrosion pits are formed in specimens after the immersion in 3.5 wt% NaCl (Figs. 6 and 7). Places privileged to creation of corrosion pits are aggregations of non-metallic inclusions (Fig. 6). On the steel surface the cracked passive layer and symptoms of intergranular corrosion running along the grain boundaries can be also observed (Fig. 7).

Figure 5. Micropores and corrosion pits on the X5CrNi18-8 steel surface in a delivery state after immersion tests in 3.5N H_2SO_4 solution.

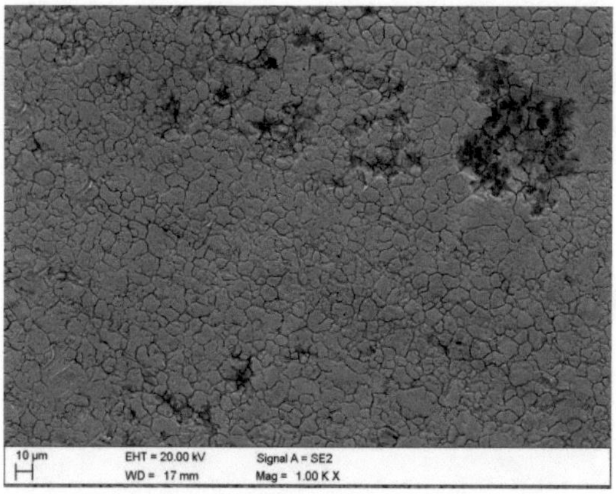

Figure 6. Corrosion pits on the X5CrNi18-8 steel surface in a delivery state after immersion tests in 3.5 wt% NaCl solution.

After cold working with the deformation amount of 50% and subsequent immersion tests in 3.5N H_2SO_4 the steel shows a meaningful percentage mass decrement up to 67% and about two orders of magnitude lower in chloride solution (Table 3). Cold deformation causes the increase of the mass loss both in acidic and chloride media in comparison with the specimens investigated in the undeformed state (Table 2). In the case of the steel immersed in 3.5N H_2SO_4 the difference between the mass loss of non-deformed and plastically deformed samples is nearly twice. The mass loss of the cold deformed samples dipped in chloride solution is three times higher. The increasing of mass loss of deformed samples is probably related to the occurrence of strain-induced α'-martensite (Fig. 8).

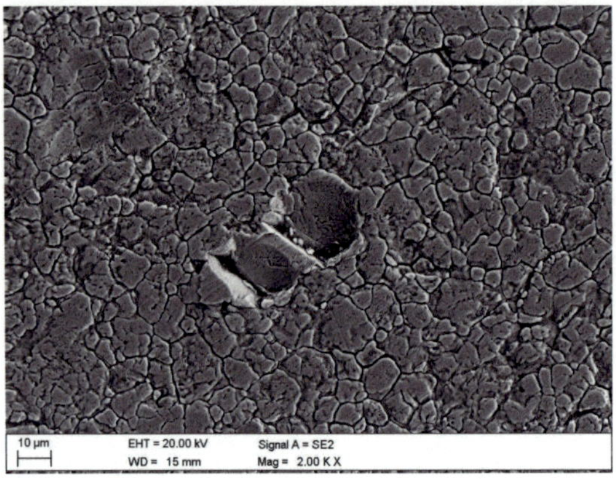

Figure 7. Corrosion pits and the cracked surface layer of the X5CrNi18-8 steel in a delivery state after immersion tests in 3.5 wt% NaCl solution.

The morphology of a surface of the cold-deformed X5CrNi18-8 steel samples after exposition in chloride and acidic media showed a various character of corrosion products (Figs. 9-16). Many corrosion pits of various size can be observed after immersion in the 3.5N H_2SO_4 solution. In the specimen deformed to 40% cold reduction wide pits and micropores can be seen (Figs. 9 and 10). The amount and the size of pits are very high and they are formed along the entire surface of specimens. Privileged places for pits forming are surface concentrations of non-metallic inclusions, which are also probable places of hydrogen penetration.

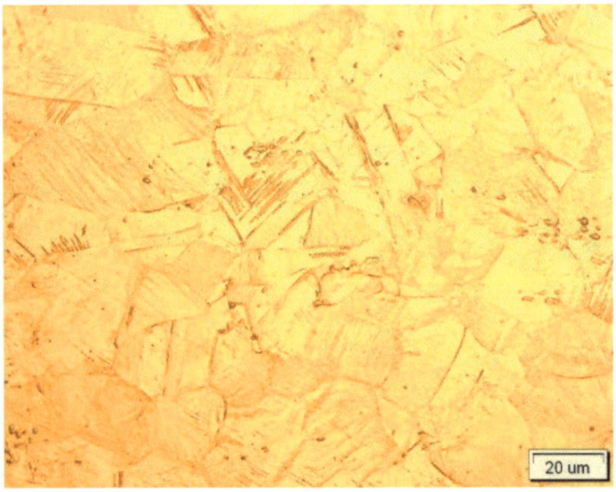

Figure 8. Elongated austenitic grains containing deformation twins and α′ martensite in the X5CrNi18-8 steel deformed to the reduction of 30%.

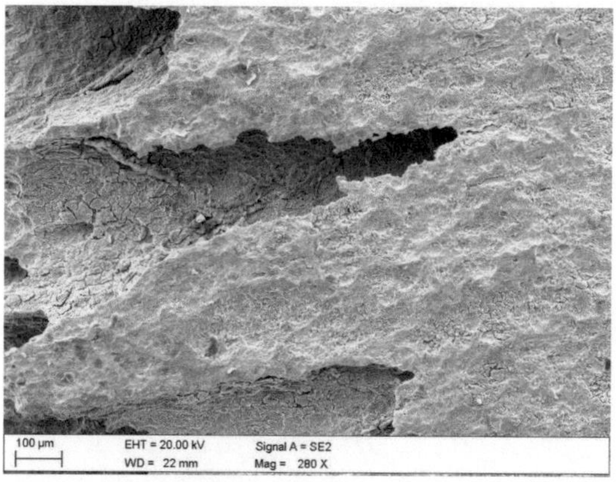

Figure 9. Numerous corrosion pits on the surface of X5CrNi18-8 steel 40% deformed and immersed in 3.5N H_2SO_4.

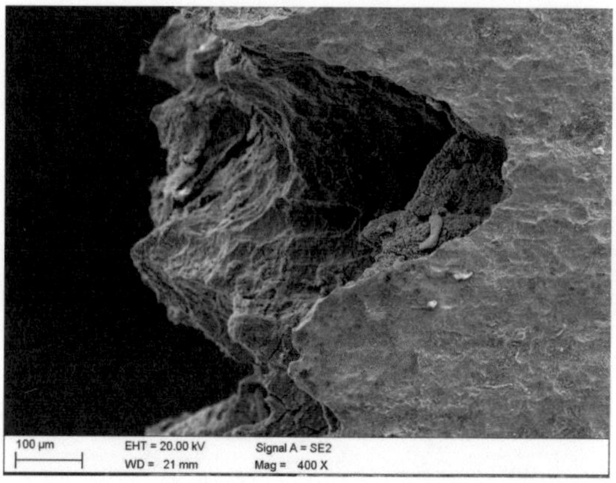

Figure 10. Corrosion pits on the surface of X5CrNi18-8 steel 40% deformed and immersed in 3.5N H_2SO_4.

Figure 11. Cracked passive layer on the surface of X5CrNi18-8 steel 40% deformed and immersed in 3.5N H_2SO_4.

Hydrogen can also penetrate deeper into the steel – probably by α' martensite laths – accumulating in a surroundings of non-metallic sulfide inclusions (Garcia et al., 2010). Usually, the places of hydrogen accumulation are non-metallic inclusions, grain boundary areas and/or twin boundaries (Abreu et al., 2007; Singh & Ray, 2007). A cracked passive layer was also observed, what could result in rapid penetration of corrosive medium into interior of the investigated steel (Fig. 11).

The surface of X5CrNi18-8 steel specimens cold rolled in a range from 10 to 70% and immersed in 3.5 wt% NaCl solution for 87 days reveals numerous pits and micropores of diversified size (Figs. 12-15). The presence of corrosion pits in a chloride medium was also confirmed by other authors (Singh et al., 2011; Xu & Hu, 2004). The pits present on the surface of deformed samples are characterized by bigger sizes compared to the pits observed on undeformed samples (Figs. 6 and 7). Damaging of a superficial layer occurred around the formed pits (Figs. 13 and 14). A cracked passive layer can be also observed. On surfaces of all the investigated samples the traces of intercrystalline corrosion running by the borders of grains and numerous pits with effects of cold rolling are visible (Figs. 12-15).

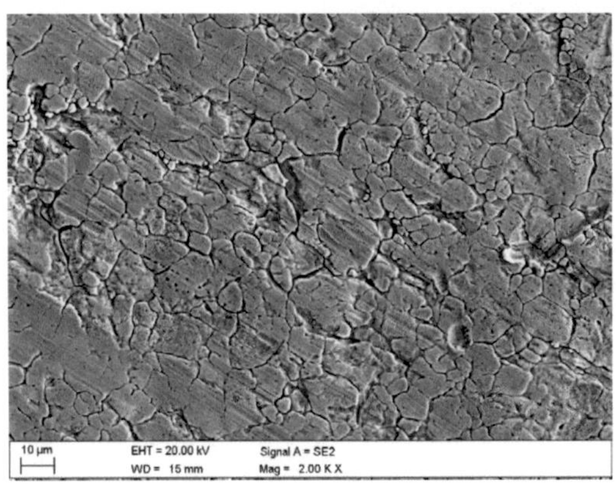

Figure 12. Numerous corrosion pits on the surface of the X5CrNi18-8 steel 10% deformed and immersed in 3.5 wt% NaCl.

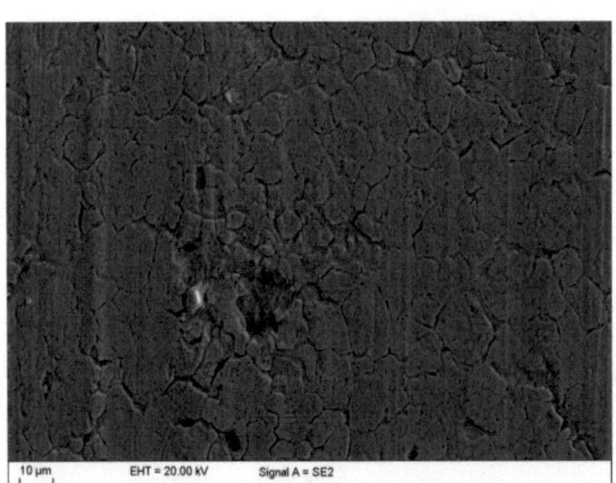

Figure 13. Corrosion pits on the surface of the X5CrNi18-8 steel 20% deformed and immersed in 3.5 wt% NaCl.

http://dx.doi.org/10.5772/53590

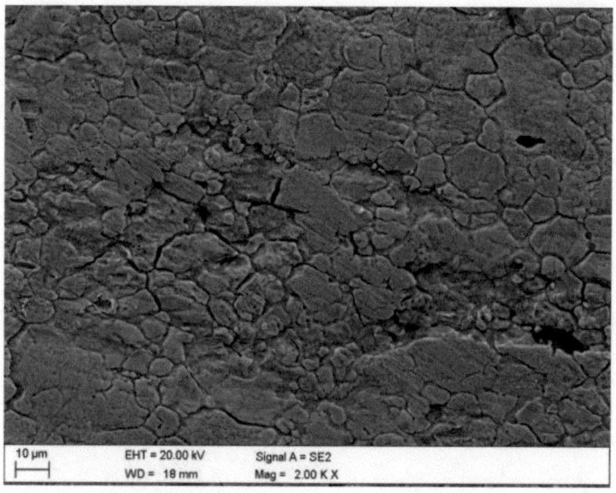

Figure 14. Corrosion pits on the surface of the X5CrNi18-8 steel 30% deformed and immersed in 3.5 wt% NaCl.

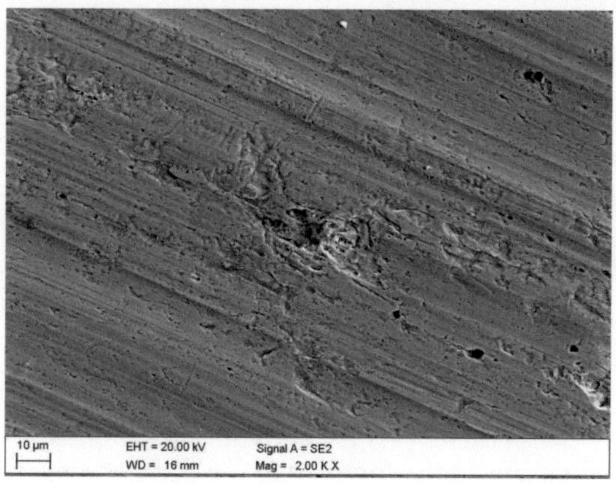

Figure 15. Corrosion pits on the surface of the X5CrNi18-8 steel 70% deformed and immersed in 3.5 wt% NaCl.

The high difference in corrosion behaviour of the stainless steel is due to different corrosion mechanisms in both environments. The big mass loss of the samples in the H_2SO_4 solution is due to the hydrogen depolarization mechanism, which is typical for corrosion in acidic media. Hydrogen depolarization is a process of reducing hydrogen ions (from the electrolyte) in cathodic areas by electrons from the metal, to gaseous hydrogen, resulting in continuous flow of electrons outer the metal and consequently the corrosion progress. Due to this process, numerous corrosion pits occur in examined steels (Figs. 5, 9 and 10). Corrosion pits occur intensively in the places containing non-metallic inclusions. They are less precious than the rest of material, fostering potential differences and galvanic cell creation. This causes the absorption of hydrogen ions, which, due to increasing pressure and temperature can recombine to a gaseous form and get out of the metal accompanying formation of corrosion pits. This process is accompanied by local cracking of corrosion products layer (Fig. 11), uncovering the metal surface and causing further penetration of the corrosive medium and the intensive corrosion progress.

The oxygen depolarization is the main mechanism of the corrosion progress in NaCl solution. In this process, oxygen included in the electrolyte is being reduced by electrons from the metal to hydroxide ions. On the surface of the alloy appears a layer of corrosion products, protecting the material before further penetration of the corrosion medium. This is why the mass loss in chloride solution is much lower compared to acidic medium (Table 2). At less corrosion-resistant places (e.g. with non-metallic inclusions) potential differences are occurring. This enables the absorption of chloride ions, which form chlorine oxides of increased solubility. It leads to local destructions of corrosion products layer and the initiation of corrosion pits. Further pit expansion is running autocatalytic.

The SEM observations of plastically deformed samples after corrosion tests permit to affirm that plastic deformation results in the significant intensification of the corrosion progress. It can be stated that increasing of cold deformation values from 10 to 70% leads to the increase of quantity, size and depth of corrosion pits fulfilling a function of a local anode.

4.1.2. Results of potentiodynamic polarization tests

Performed electrochemical analysis revealed that the open circuit potential for undeformed and cold deformed samples established itself after 60 min. The change of current density as a function of potential for the undeformed samples investigated in 3.5 wt% NaCl solution is presented in Fig. 16. The value of corrosion potential E_{cor} was equal -48 mV and the density of corrosion current i_{cor} determined basing on the Stern-Geary equation was equal 0.01 $\mu A/cm^2$. The recorded anodic polarization curve indicates the existence of a passive range. The value of the breakdown potential is equal to +360 mV. When the current density reached 3 mA/cm^2 the direction of anodic polarization of samples was changed. The change of the polarization direction caused the increase of the current density.

The corrosion potential of the X5CrNi18-8 steel cold deformed with the reduction from 10 to 70% covers the range from –91 to –51 mV (Fig. 17). The values of corrosion current density were equal to 0.02 ÷ 1.16 $\mu A/cm^2$ and the breakdown potential changed in the range from +245 to +348 mV. The similar values of the corrosion potential and corrosion current density in a chloride medium for Cr-Ni austenitic steels are reported by other authors (Rutkowska-Gorczyca et al., 2009; Zhilin et al., 2006; Kumar et al., 2007). The changes in the pitting potential, in cold deformed samples, have been reportedly attributable to stresses, strain-induced α'-martensite or dislocations. The martensite phase can not only cause a decrease in the solution resistance within the pit but also caused a fall in the polarization resistance between metal and a pit. Thus, pit propagation in the 304 stainless steel is accelerated with increasing volume fraction of martensite (Kamide et al., 1994). On the basis of the anodic polarization curves obtained for the investigated X5CrNi18-8 steel it has been found that with increasing the degree of plastic deformation within the range from 10 to 70% the steel is characterized by the lower values of corrosion potential as well as higher values of corrosion current density (Fig. 17), what proves the intensification of the corrosion progress due to cold working.

http://dx.doi.org/10.5772/53590

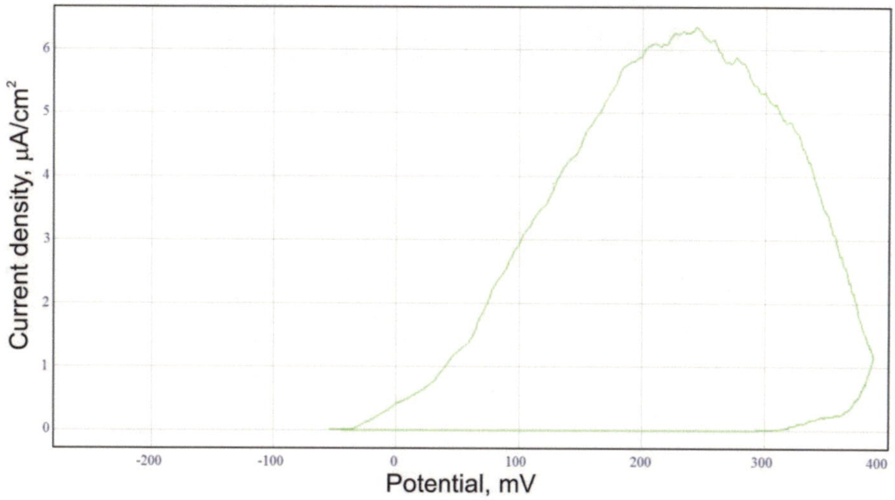

Figure 16. Anodic polarization curve registered for the X5CrNi18-8 steel in a delivery state in 3.5 wt% NaCl.

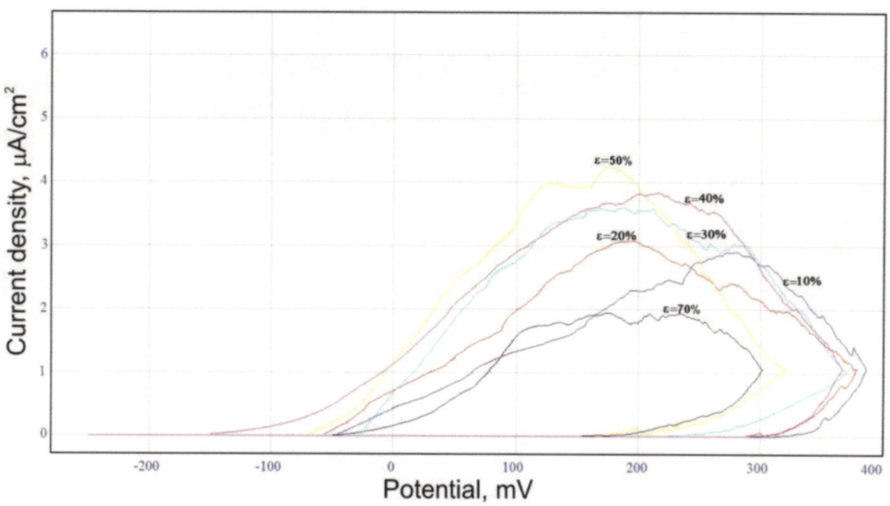

Figure 17. Anodic polarization curves registered for the cold-deformed samples of X5CrNi18-8 steel in 3.5 wt% NaCl.

SEM observations of a sample surface after the electrochemical corrosion tests in 3.5 wt% NaCl allowed to evaluate the type and degree of corrosion damages. The surface of the X5CrNi18-8 steel in a delivery state is characterized by relatively small corrosion pits and a cracked surface layer (Fig. 18). It is assumed that the surface of the examined samples was subjected to the greatest corrosion attack in places where the local breakdown of the passivation oxide layer in the presence of aggressive anions of the environment has occurred.

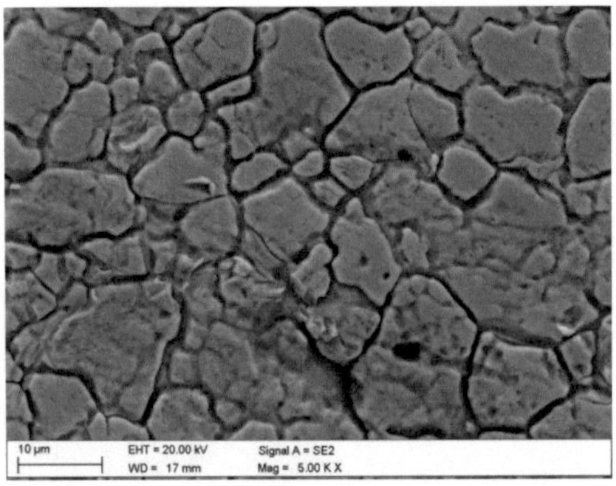

Figure 18. Corrosion pits on the surface of X5CrNi18-8 steel in a delivery state after electrochemical corrosion tests in 3.5 wt% NaCl solution.

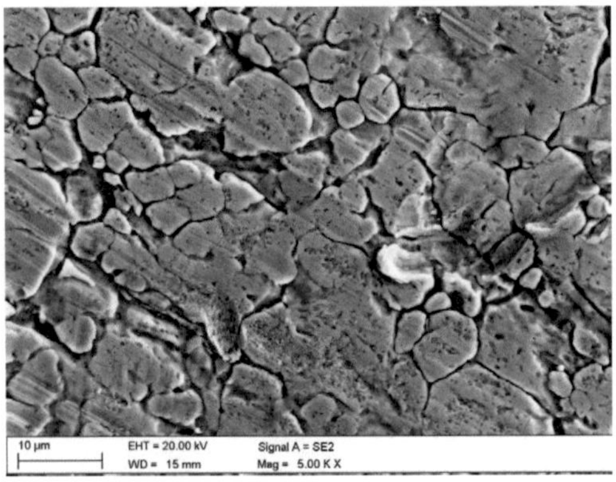

Figure 19. Corrosion pits and the partially cracked surface layer of the X5CrNi18-8 steel 10% deformed and electrochemically tested in 3.5 wt% NaCl.

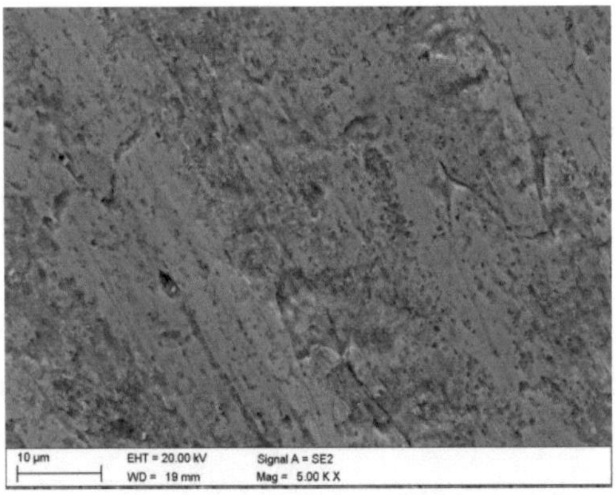

Figure 20. Numerous corrosion pits on the surface of the X5CrNi18-8 steel 70% deformed and electrochemically tested in 3.5 wt% NaCl.

The surfaces of the electrochemically tested X5CrNi18-8 steel specimens after cold rolling in a range from 10 to 70% show also numerous pits, micropores and cracks of a surface layer (Figs. 19 and 20). The pits present on the surface of deformed samples are characterized by bigger sizes compared to the pits observed in undeformed samples (Fig. 18). Obtained microstructural results correspond well with those registered for immersion tests.

4.2. High-Mn austenitic steel

The results of immersion tests for the high-Mn steel are given in Table 3. After 4 days immersion in 1N H_2SO_4 the investigated steel shows a significant mass decrement equal to about 38%. Mass loss of samples dipped in a chloride solution is about 100 times lower. The results are similar to these obtained for the stainless steel (Table 2), however, the high-Mn steel corroded to the same extent at much shorter time and under softer acid solution conditions. Therefore, the real corrosion progress of the high manganese steel is much faster compared to the Cr-Ni steel.

X6MnSiAl26-3-3 steel	**Corrosion medium**	
State	1N H_2SO_4	3.5 wt% NaCl
non-deformed	38.4 ± 5.2	0.40 ± 0.03
cold deformed	47.5 ± 1.6	0.33 ± 0.01

Table 3. Mean percentage mass loss of samples in an initial state and cold-deformed by bending after the immersion tests, %

Numerous corrosion pits along the whole specimen surface can be observed after immersion of the X6MnSiA126-3-3 steel in 1N H_2SO_4 (Fig. 21). Slightly smaller corrosion pits are formed in specimens investigated in 3.5 wt% NaCl. The surface layer of specimens dipped in the acidic medium is characterized by the presence of many cracks and craters formed due to corrosion pitting (Fig. 22). The cracks in the neighbourhood of non-metallic inclusions are also visible in Figure 23 showing the surface of a sample dipped in the chloride solution. Under these conditions, a layer of corrosion products is forming, protecting the steel against further penetration of corrosive medium.

Similar surface defects were also revealed on surfaces of plastically deformed samples. Privileged places for forming of corrosion pits and surface cracks are concentrations of non-metallic inclusions, which are also potential places of hydrogen penetration. Figure 24 presents the plastically deformed austenitic grains containing deformation twins and elongated non-metallic inclusions. It was observed that hydrogen can penetrate into the steel up to a depth of about 0.3 mm, accumulating usually in a surroundings of elongated sulfide inclusions. Hydrogen failures can be usually observed at non-metallic inclusions and in grain boundary areas (Fig. 24).

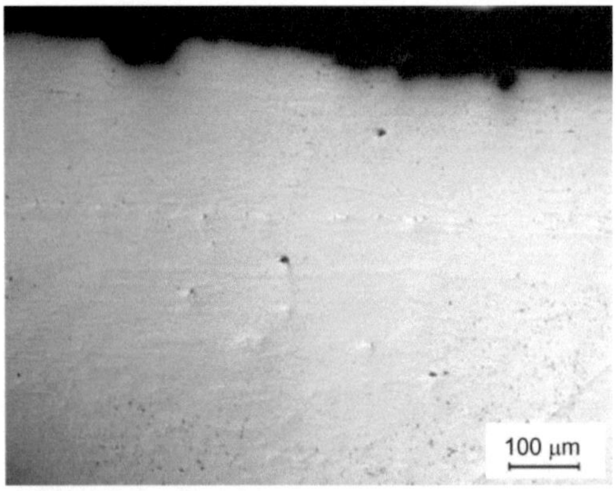

Figure 21. Corrosion pits and non-metallic inclusions in the X6MnSiAl26-3-3 steel after the immersion test in 1N H_2SO_4.

Figure 22. Numerous cracks and craters formed due to corrosion pitting on the surface of the X6MnSiAl26-3-3 steel after the immersion test in 1N H_2SO_4.

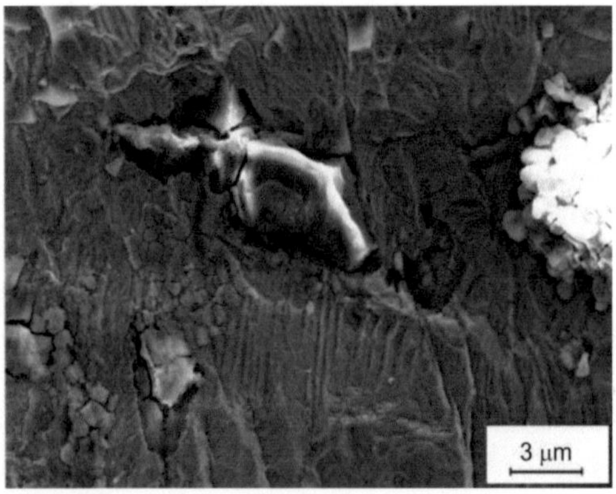

Figure 23. A layer of corrosion products containing surface cracks in the X6MnSiAl26-3-3 steel after the immersion test in 3.5 wt% NaCl.

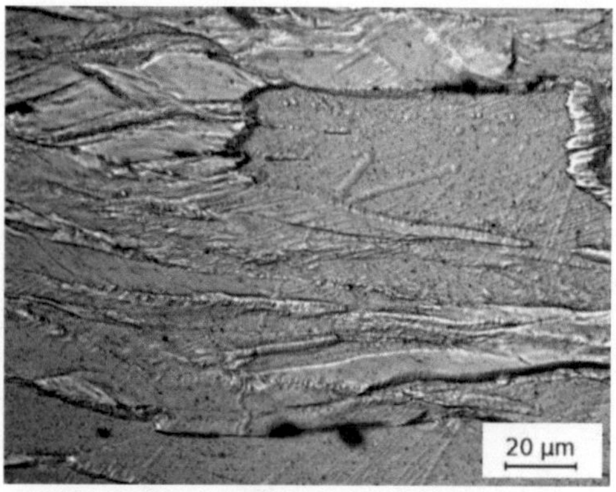

Figure 24. Austenitic grains containing deformation twins, elongated sulfide inclusions and hydrogen failures in the X6MnSiAl26-3-3 steel after bending and immersion in 1N H_2SO_4.

Figure 25. Cracked layer of corrosion products with banding-like arrangement in the X6MnSiAl26-3-3 steel after bending and immersion in 1N H_2SO_4.

Figure 26. Cracked layer of corrosion products and deep corrosion decrements in the X6MnSiAl26-3-3 steel after bending and immersion in 1N H_2SO_4.

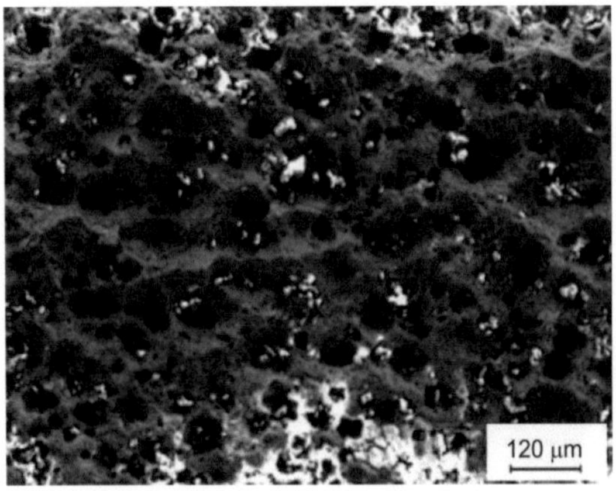

Figure 27. Numerous craters formed due to corrosion pitting and probable hydrogen impact in the X6MnSiAl26-3-3 steel after bending and immersion in 1N H_2SO_4.

The micrographs in Figures 25 and 26 reveal deep corrosion decrements and band-arranged corrosion products in a surface area. The corrosion products layer is not continuous and has many cracks (Fig. 26). Besides corrosion products, numerous craters formed due to intensive corrosion pitting and probably as a result of hydrogen impact are characteristic (Fig. 27). Craters forming is accompanied by local cracking of corrosion products layer (Fig. 26), uncovering the metal surface and causing further penetration of the corrosive medium and finally the intensive progress of general and pitting corrosion.

Generally Hydrogen Induced Cracking (HIC) is not a problem in austenitic steels because of the relatively low diffusion coefficient of H in a FCC lattice (Kumar & Balasubramaniam, 1997). However, enhanced permeation of hydrogen was observed in cold worked austenitic steels by Kumar & Balasubramaniam (1997), what was attributed to a strain-induced martensitic transformation leading to promote hydrogen diffusion as the diffusivity is much higher in the bcc martensite lattice. Additionally, hydrogen mobility is enhanced by the presence of high-dislocation density due to cold working (Ćwiek, 2010). The hydrogen induced surface cracking at the high hydrogen concentration places, i.e. grain and twin boundaries, ε/γ interface was also observed in stainless steels during hydrogen effusion from the supersaturated sites (Yang & Luo, 2000). Especially privileged to surface hydrogen accumulation is ε martensite. It was observed in a X5MnSiAl25-4-2 high-Mn steel containing lamellar plates of ε martensite (Grajcar et al., 2010). Atomic hydrogen absorbed in the surface area penetrates the steel and accumulates in places with non-metallic inclusions, lamellar precipitations of the second phase, microcracks and other structural defects, where convenient conditions for recombining of atomic hydrogen to molecular H_2 exist. The recombination of atomic hydrogen to molecular state is a very exothermic reaction, which provides a pressure increase in formed H_2 bubbles as well as nucleation and growth of microcracks in a surface region of the sample. In the investigated X6MnSiAl26-3-3 steel the evident microcraks associated with HIC were not revealed but observed corrosion damages in a surface area (Figs. 24-27) are a result of a combined contribution of general corrosion, pitting and hydrogen impact.

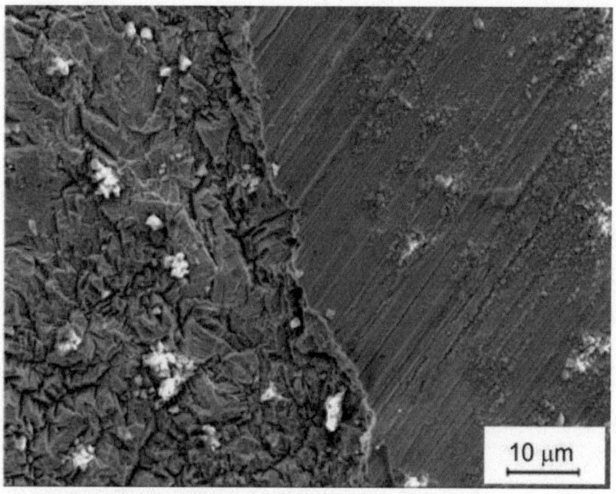

Figure 28. The boundary between undipped and dipped parts of the sample showing a scaled and cracked layer of corrosion products in the X6MnSiAl26-3-3 steel after bending and immersion in 3.5 wt% NaCl.

A layer of corrosion products protecting the metal against continuous penetration of a corrosive medium is forming on the surface of specimens dipped in a chloride solution (Fig. 28). The formed scaled layer strongly adheres to the steel though numerous surface cracks occurring. Corrosion cracks were not observed as distinct from a steel containing ε martensite plates (Grajcar, 2012). The created surface layer of corrosion products protects effectively the base metal against chloride medium even after cold deformation because the mass decrement is comparable to that obtained for non-deformed samples (Table 3). In turn, cold deformation rises slightly the mass loss in the acidic medium in comparison with the specimens investigated in an initial state (Table 3). The effect of cold deformation on the acceleration of mass losses was much higher for the stainless steel (Table 2) probably due to the occurrence of α' martensite forming a local corrosive galvanic cell with the austenitic matrix. The enhancement of the corrosion progress in the high-Mn steel immersed in the acidic solution can be only related to annealing twins showing a different potential than the matrix (Ghayad et al., 2006; Mazancova et al., 2010).

5. Summary

The dynamic development of technology creates the necessity for producing steels combining high strength, toughness and ductile properties with high corrosion resistance. The austenitic alloys possess an exceptional balance of strength and ductility as well as the high ability to further rise their strength during technological forming or cold rolling. These properties decide about their wide application in the chemical, machinery, food, automotive, nuclear and shipbuilding industries. Although austenitic stainless steels are characterized by excellent resistance to general corrosion, they are susceptible to the localized corrosive attacks, such as pitting corrosion, intergranular corrosion and stress corrosion cracking (SCC) in chloride and sulfuric environments. The high-manganese austenitic alloys can be also an option for some applications but generally their corrosion resistance is much lower.

The results presented in this chapter focused on the evaluation of corrosion behaviour of plastically deformed Cr-Ni and high-Mn steels in acidic and chloride containing media. The results of immersion and potentiodynamic tests as well as microstructural studies prove that both examined steels are susceptible to corrosion attacks. However, the corrosion progress in the Cr-Ni stainless steel requires stronger corrosive reagents and longer exposition time in corrosive environments compared to the steel alloyed with manganese.

The mass loss of the X5CrNi18-8 steel immersed in 3.5N H_2SO_4 is equal to about 40% and is over two orders of magnitude higher compared to the specimens dipped in the chloride solution. The similar difference in the corrosion progress was obtained for the high-Mn steel, though its real corrosion rate (taking into account test conditions) is much higher than that of the stainless steel. The fast progress of mass loss in the acidic solution is a result of the hydrogen depolarization, which is a typical corrosion mechanism under such conditions. Due to this process, numerous corrosion pits are formed and local cracking of corrosion products layer occurs resulting in penetration of the corrosive medium and finally leading to the intensive corrosion progress. In turn, the oxygen depolarization process results in formation of a layer of corrosion products on surfaces of the steels examined in the chloride medium limiting effectively a corrosion progress.

The cold working intensifies the corrosion progress especially in the Cr-Ni steel as a result of formation of the strain-induced α' martensite. The percentage mass decrements are near twice and three times higher correspondingly for the acidic and chloride solutions. The increase of the degree of plastic deformation from 10% up to 70% results in lowering the corrosion potential as well as higher values of corrosion current density. The pit forming and propagation in the X5CrNi18-8 steel are accelerated with increasing volume fractions of martensite. The corrosion rate and subsequent mass loss of the cold deformed high-Mn steel increase only slightly probably due to deformation twins forming.

In general, the low corrosion resistance of high-manganese steel is from the fact, that Mn forms unstable manganese oxide due to low passivity coefficient and hence reduces their electrochemical corrosion resistance (Kannan et al., 2008). It leads consequently to the high dissolution rate of Mn and Fe both in H_2SO_4 and NaCl solutions (Ghayad et al., 2006; Hamada, 2007; Kannan et al., 2008; Zhang & Zhu, 1999). It seems that high-Mn steels can replace conventional austenitic stainless steels only partially in non-critical applications. Their corrosion resistance can be improved by surface engineering, i.e. zinc coatings or Cr alloying (Hamada, 2007; Mujica Roncery et al., 2010).

Acknowledgement

The work was partially financially supported by the NCN – The National Science Centre (grant No. 2632/B/T02/2011/40).

Author details

Wojciech Ozgowicz, Agnieszka Kurc-Lisiecka and Adam Grajcar

Silesian University of Technology, Gliwice, Poland

References

[1] Abreu, H.; Carvalho, S.; Neto, P.; Santos, R.; Freire, V.; Silva, P. & Tavares, S. (2007). Deformation induced martensite in an AISI 301LN stainless steel: Characterization and influence on pitting corrosion resistance. *Materials Research*, Vol.10, (2007), pp. 359-366

[2] Altstetter, C.J.; Bentley, A.P.; Fourine, J.W. & Kirkbridge, A.N. (1986). Processing and properties of Fe-Mn-Al alloys. *Materials Science and Engineering A,* Vol.82 (1986), pp. 13-25

[3] Azambuja, D.S.; Martini, E.M. & Müller, I.L. (2003). Corrosion behaviour of iron and AISI 304 stainless steel in tungstate aqueous solutions containing chloride. *Journal of the Brazilian Chemical Society,* Vol.14, No.4, (2003), pp. 570-576

[4] Barbucci, A.; Cerisola, G. & Cabot, P.L. (2002). Effect of cold working in the passive behavior of 304 stainless steel in sulfate media. *Journal of Electrochemical Society,* Vol. 149, (2002), pp. 534-542

[5] Baszkiewicz, J. & Kamiński, M. (1997). *Fundamentals of materials corrosion,* The Warsaw University of Technology Publishers, Warsaw, Poland

[6] Briant, C.L. (1982). Effect of nitrogen and cold work on the sensitization of austenitic stainless steels. *Electric Power Research Institute Report* EPRI-NP-2457, Palo Alto, California

[7] Ćwiek, J. (2009). Hydrogen degradation of high-strength steel. *Journal of Achievements in Materials and Manufacturing Engineering,* Vol.37, (2009), pp. 193-212

[8] Ćwiek, J. (2010). Prevention methods against hydrogen degradation of steel. *Journal of Achievements in Materials and Manufacturing Engineering,* Vol.43, No.1, (2010), pp. 214-221

[9] De Cooman, B.C.; Chin, K. & Kim, J. (2011). High Mn TWIP steels for automotive applications, In: *New Trends and Developments in Automotive System Engineering,* M. Chiaberge, (Ed.), pp. 101-128, InTech, ISBN 978-953-307-517-4, Rijeka, Croatia

[10] Fang, Z.; Wu, Y.S.; Zhang, L. & Li, J. (1997). Effect of deformation induced martensite on electrochemical behaviors of type 304 stainless steel in the active state. *Corrosion Science and Protection Technology,* Vol.9, No.1, (1997), pp. 75-83

[11] Frankel, G.S. (1998). Pitting corrosion of metals. A review of the critical factors. *Journal of the Electrochemical Society*, Vol.145, No.6, (1998), pp. 2186-2198

[12] Frommeyer, G. & Bruex, U. (2006). Microstructures and mechanical properties of high-strength Fe-Mn-Al-C light-weight TRIPLEX steels. *Steel Research International*, Vol.77, No.9-10, (2006), pp. 627-633

[13] Frommeyer, G.; Bruex, U. & Neumann, P. (2003). Supra-ductile and high-strength manganese-TRIP/TWIP steels for high energy absorption purposes. *ISIJ International*, Vol.43, No.3, (2003), pp. 438-446

[14] Fu, Y.; Wu, X.; Han, E.H.; Ke, W.; Yang, K. & Jiang, Z. (2009). Effects of cold work and sensitization treatment on the corrosion resistance of high nitrogen stainless steel in chloride solutions. *Electrochimica Acta*, Vol.54, (2009), pp. 1618-1629

[15] Garcia, E.G.; Paniagua, F.A.; Herrera-Hernández, H.; Juárez Garcia, J.M.; Pardavé M.E. & Romo, M.A. (2010). Electrochemical and microscopy study of localized corrosion on a sensitized stainless steel AISI 304. *ESC Transactions*, Vol.29, No.1, (2010), pp. 93-102

[16] Ghayad, I.M.; Hamada, A.S.; Girgis, N.N. & Ghanem, W.A. (2006). Effect of cold working on the aging and corrosion behaviour of Fe-Mn-Al stainless steel. *Steel Grips*, Vol.4, No.2, (2006), pp. 133-137

[17] Graessel, O.; Krueger, L.; Frommeyer, G. & Meyer, L.W. (2000). High strength Fe-Mn-(Al, Si) TRIP/TWIP steels development – properties – application. *International Journal of Plasticity*, Vol.16, (2000), pp. 1391-1409

[18] Grajcar, A. (2012). Corrosion resistance of high-Mn austenitic steels for the automotive industry, In: *Corrosion Resistance*, H. Shih, (Ed.), pp. 353-376, InTech, ISBN 978-953-51-0467-4, Rijeka, Croatia

[19] Grajcar, A.; Kołodziej, S. & Krukiewicz, W. (2010a). Corrosion resistance of high-manganese austenitic steels. *Archives of Materials Science and Engineering*, Vol.41, No. 2, (2010), pp. 77-84

[20] Grajcar, A.; Krukiewicz, W. & Kołodziej, S. (2010b). Corrosion behaviour of plastically deformed high-Mn austenitic steels. *Journal of Achievements in Materials and Manufacturing Engineering*, Vol.43, No.1, (2010), pp. 228-235

[21] Grajcar, A.; Opiela, M. & Fojt-Dymara, G. (2009). The influence of hot-working conditions on a structure of high-manganese steel. *Archives of Civil and Mechanical Engineering*, Vol.9, No.3, (2009), pp. 49-58

[22] Hamada, A.S. (2007). *Manufacturing, mechanical properties and corrosion behaviour of high-Mn TWIP steels*. Acta Universitatis Ouluensis C281, ISBN 978-951-42-8583-7, Oulu, Finland

[23] Huang, B.X.; Wang, X.D.; Rong, Y.H.; Wang, L. & Jin, L. (2006). Mechanical behavior and martensitic transformation of an Fe-Mn-Si-Al-Nb alloy. *Materials Science and Engineering A,* Vol.438-440, (2006), pp. 306-313

[24] International Iron & Steel Institute (September 2006). Advanced High Strength Steel (AHSS) Application Guidelines – version 3, Available from http://worldautosteel.org

[25] Jimenez, J.A. & Frommeyer, G. (2010). Microstructure and texture evolution in a high manganese austenitic steel during tensile test. *Materials Science Forum,* Vol.638-642, (2010), pp. 3272-3277

[26] Kamide, H.; Fujitsuka, K. & Tanaka, Y. (1994). Effect of carbon content on dissolution rate of ′ martensite and 304 stainless steel in a H_2SO_4-NaCl solution. *Journal of the Japan Institute of Metals,* Vol.58, (1994), pp. 1414-1419

[27] Kannan, M.B.; Raman, R.K.S., & Khoddam, S. (2008). Comparative studies on the corrosion properties of a Fe-Mn-Al-Si steel and an interstitial-free steel. *Corrosion Science,* Vol.50, (2008), pp. 2879-2884

[28] Khatak, H.S. & Raj, B. (2002). *Corrosion of austenitic stainless steel: mechanism, mitigation and monitoring,* Woodhead Publishing, ISBN 1-85573-613-6, London, UK

[29] Kumar, B.R.; Mahato, B. & Singh, R. (2007). Influence of cold-worked structure on electrochemical properties of austenitic stainless steels. *Metallurgical and Materials Transactions,* Vol.38A, (2007), pp. 2085-2094

[30] Kumar, B.R.; Singh, R.; Mahato, B.; De, P.K.; Bandyopadhyay, N.R. & Battacharya, D.K. (2005). Effect of texture on corrosion behavior of AISI 304L stainless steels. *Materials Characterization,* Vol.54, (2005), pp. 141-147

[31] Kumar, P. & Balasubramaniam, R. (1997). Determination of hydrogen diffusivity in austenitic stainless steels by subscale microhardness profiling. *Journal of Alloys and Compounds,* Vol.255, (1997), pp. 130-134

[32] Kurc, A.; Kciuk, M. & Basiaga, M. (2010). Influence of cold rolling on the corrosion resistance of austenitic steel. *Journal of Achievements in Materials and Manufacturing Engineering,* Vol.38, No.2, (2010), pp. 154-162

[33] Lee, T.H.; Oh, C.S.; Kim, S.J. & Takaki, S. (2007). Deformation twinning in high-nitrogen austenitic stainless steel. *Acta Materialia,* Vol.55, (2007), pp. 3649-3662

[34] Lovicu, G.; Barloscio, M.; Botaazzi, M.; D'Aiuto, F.; De Sanctis, M.; Dimatteo, A.; Federici, C.; Maggi, S.; Santus, C. & Valentini, R. (2010). Hydrogen embrittlement of advanced high strength steels for automotive use. *Proceedings of International Conference on Super-High Strength Steels,* pp. 1-13, Peschiera del Garda, Italy, October 17-20, 2010

[35] Mansur, L.K. & Lee, E.H. (1990). A mechanism of swelling suppression in cold-worked phosphorous modified stainless steels, *Philosophical Magazine A,* Vol. 61, (1990), pp. 733-749

[36] Mazancova, E.; Kozelsky, P. & Schindler, I. (2010). The TWIP alloys resistance in some corrosion reagents. *Proceedings of International Conference METAL*, pp. 1-6, Roznov pod Radhostem, Czech Republic, May 18-20, 2010

[37] Mujica Roncery, L.; Weber, S. & Theisen, W. (2010). Development of Mn-Cr-(C-N) corrosion resistant twinning induced plasticity steels: thermodynamic and diffusion calculations, production and characterization. *Metallurgical and Materials Transactions A*, Vol.41A, No.10, (2010), pp. 2471-2479

[38] Ningshen, S. & Kamachi Mudali, U. (2010). Pitting and intergranular corrosion resistance of AISI type 301LN stainless steels. *Journal of Materials Engineering and Performance*, Vol.19, No.2, (2010), pp. 274–281

[39] Oh, Y.J. & Hong, J.H. (2000). Nitrogen effect on precipitation and sensitization in cold-worked type 316L(N) stainless steels. *Journal of Nuclear Materials*, Vol.278, (2000), pp. 242-250

[40] Opiela, M.; Grajcar, A. & Krukiewicz, W. (2009). Corrosion behaviour of Fe-Mn-Si-Al. *Journal of Achievements in Materials and Manufacturing Engineering*, Vol.33, No.2, (2009), pp. 159-165

[41] Osawa, M & Hasegawa, M. (1981). Stress corrosion cracking of hydrogen-containing austenitic stainless steel in H_2SO_4 –NaCl solution. *Transactions of the Iron and Steel Institute of Japan*, Vol.21, (1981), pp. 464-468

[42] Otero, E.; Pardo, A.; Sáenz, E.; Utrilla, V. & Pérez F. (1995). Intergranular corrosion behaviour of a new austenitic stainless steel low in nickel. *Canadian Metallurgical Quarterly*, Vol.34, (1995), pp. 135-141

[43] Pardo, A.; Merino, M.C.; Carboneras, M. & Coy A.E. (2007). Pitting corrosion behaviour of austenitic stainless steels with Cu and Sn additions. *Corrosion Science*, Vol.49, (2007), pp. 510-525

[44] Peguet, L.; Malki, B. & Baroux, B. (2007). Influence of cold working on the pitting corrosion resistance of stainless steels. *Corrosion Science*, Vol.49, (2007), pp. 1933-1948

[45] Qiao L.J. & Luo J.L. (1998). Hydrogen-facilitated anodic dissolution of austenitic stainless steels. *Corrosion Science*, Vol. 54, No. 4, (1998), pp. 281-288

[46] Rutkowska-Gorczyca, M.; Podrez-Radziszewska, M. & Kajtoch, J. (2009). Influence of cold working process on the corrosion resistance of steel 316L. *SIM XXXVII*, Kraków-Krynica, (2009), pp. 319-323

[47] Shin, S.Y.; Hong, S.; Kim, H.S.; Lee, S. & Kim, N.J. (2010). Tensile properties and cup formability of high Mn and Al-added TWIP steels. *Proceedings of International Conference on Super-High Strength Steels*, pp. 1-9, Peschiera del Garda, Italy, October 17-20, 2010

[48] Sing M.K. & Kumar A. (2011). Environmental corrosion studies of cold rolled austenitic stainless steel. *International Journal of Advanced Scientific and Technical Research,* Vol.2, No.1, (2011), pp.470-482

[49] Sing V.B. & Ray, M. (2007). Effect of H_2SO_4 addition on the corrosion behaviour of AISI 304 austenitic stainless steel in methanol-HCl solution. *International Journal of Electrochemical Science,* Vol.2, (2007), pp.329-340

[50] Sunada, S.; Kariba, M.; Majami, K. & Sugimoto, K. (2006). Influence of concentration of H_2SO_4 and NaCl on stress corrosion cracking of SUS304 stainless steel in H_2SO_4-NaCl aqueous solution. *Materials Transactions,* Vol. 47, No.2, (2006), pp.364-370

[51] Sunada, S.; Nakamura, N.; Kawase, H.; Notoya, H.; Sanuki, S. & Arai, K. (1991). Effect of deformation-induced martensite on the general corrosion of SUS304 stainless steel in H_2SO_4-NaCl solution. *Journal of the Japan Institute of Metals,* Vol. 55, No.6, (1991), pp.1078-1089

[52] Štefec, R. & Franz, F. (1978). A study of the pitting corrosion of cold-worked stainless steel. *Corrosion Science,* Vol.18, No.2, (1987), pp.161-168

[53] Xu, C. & Hu, G. (2004). Effect of deformation-induced martensite on pit propagation behavior of 304 stainless steel. *Anti-Corrosion Methods and Materials,* Vol.51, (2004), pp. 381-388

[54] Yang, Q. & Luo, J.L. (2000). Martensite transformation and surface cracking of hydrogen charged and outgassed type 304 stainless steel. *Materials Science and Engineering A,* Vol.288, (2000), pp. 75-83

[55] Zhang, Y.S. & Zhu, X.M. (1999). Electrochemical polarization and passive film analysis of austenitic Fe-Mn-Al steels in aqueous solutions. *Corrosion Science,* Vol.41, (1999), pp. 1817-1833

[56] Zhilin, L.; Wei, L. & Juncai Q. (2006). The effect of electrochemically induced annealing on the pitting resistance of metastable austenitic stainless steel. *Metallurgical and Materials Transactions,* Vol.37A, No.2, (2006), pp. 435-439

Chapter 5

Stress Corrosion Cracking of Ductile Ni-Resist Irons and Stainless Steels

Osama Abuzeid, Mohamed Abou Zour,
Ahmed Aljoboury and Yahya Alzafin

Additional information is available at the end of the chapter

http://dx.doi.org/10.5772/52318

1. Introduction

Potable water, in the Arabian Gulf and many other regions around the world, is mainly produced by desalinating seawater. Multi-stage flashing chambers (MSF) desalination plants are reported to account for producing about 85% of the desalinated water in the world [1]. In these plants, large heavy duty vertical brine circulation pumps (BCP) are used. Brine is a very corrosive environment rich in chlorides. During their operation, BCP are subjected to continuous hydraulic and mechanical loading while handling a very corrosive environment with high chloride content. These operating conditions are enough to initiate stress corrosion cracking SCC. Failure of these critical pumps would result in costly shut downs of the desalination plant and thus affecting plant reliability and availability. The rotating parts of brine circulation pumps are usually made out of austenitic stainless steels or duplex stainless steels, whereas, pressure casings had been made out of ductile Ni-resist irons (DNI) at least till the 1990's, beyond which more resistant materials have been the preferred choice of construction; e.g. Duplex Stainless Steels. It is however a fact that many of the pumps in operation are still made of DNI which are highly alloyed class of cast irons. The main alloying element in DNI is Nickel and its content varies between 18% and 22%, giving its austenitic microstructure and its desirable corrosion resistance properties. Their microstructure is characterized by uniformly distributed nodular graphite in an austenitic matrix which also contains carbide areas. DNI have also good erosive wear resistance, good machineability, castability and controlled expansion.

Meanwhile, austenitic and duplex SS materials are gaining more popularity as pump casings materials than DNI in brine environment. In addition to cast stainless steels, original pump manufacturers sometimes use welded construction of wrought stainless steels to build other related components such as column pipes and discharge elbow piece [2]. Due to the difference in expensive alloying, and apparently higher demand in many appliactions, austenitic SS is cheaper than duplex and superduplex SS. However, the mechanical, corrosion and SCC properties of duplex and superduplex SS are superior to that of austenitic SS. Therefore, economically, the idea of using chemical corrosion inhibitors to enhance the SCC resistance of the austenitic SS, is appealing, non famous and worth looking at.

In this chapter, design and construction of an SCC testing rig and testing method are described. A comparison between two types of widely used DNI in building BCP is carried out. Mechanical, metallurgical, electrochemical and SCC test results are reported. This is followed by presenting similar test results for two types of stainless steel that started replacing DNI in manufacturing pump casings [3]. Wrought stainless steel samples of the two types are used in performing the comparison between the behavior of the two types. Finally an attempt is reported to improve the immunity of the cheaper austenitic stainless steel through using chemical treatment via one proven performance corrosion inhibitor.

2. SCC testing rig and method

Fig. 1 shows a photograph for a constructed SCC test rig [2]. The rig is designed to simulate real service conditions in a desalination plant. It comprises a proof ring containing a testing chamber, a constant load tightening screw system, brine container with heating plate and other attached accessories such as electrodes, wiring to an ACM potentiostat, a computer, a dial indicator to monitor ring deflection during SCC testing and a web monitoring camera. The proof ring is made from a duplex stainless steel and is used to control the load on the SCC test samples. It is welded to upper and lower bosses. Both bosses were drilled through the ring. The lower boss is used to fix the ring in place, whereas, the upper boss is used in mounting the tightening screw loading system. A 100 kN (MTS) tension compression testing machine can be used to calibrate the bossed ring to convert its axial deflection into axial load on the SCC sample. The SCC testing chamber which is made from transparent acrylic tube is used to accommodate the SCC testing sample, hot brine, electrodes and a thermocouple. A top and bottom Teflon covers, each with an O ring seal, are used together with the acrylic tube to form the testing chamber. Four holes are drilled in the top cover to fit the working electrode (SCC sample), auxiliary electrode, reference electrode, and a thermocouple. A constant tensile load mechanism consisting of a tightening screw and nut system made from 316L stainless steel is used to pull up the tested sample. A tightening nut is used to maintain the ring deflection to a level corresponding to the required tensile load as given by the calibration data of the proof ring.

To ensure that only tensile stresses are transmitted to the sample without any torsion shear stresses, a properly devised stressing jig can be used. Inside the testing chamber, samples are subjected also to circulated hot brine of controlled temperature between 55 °C and 60 °C.

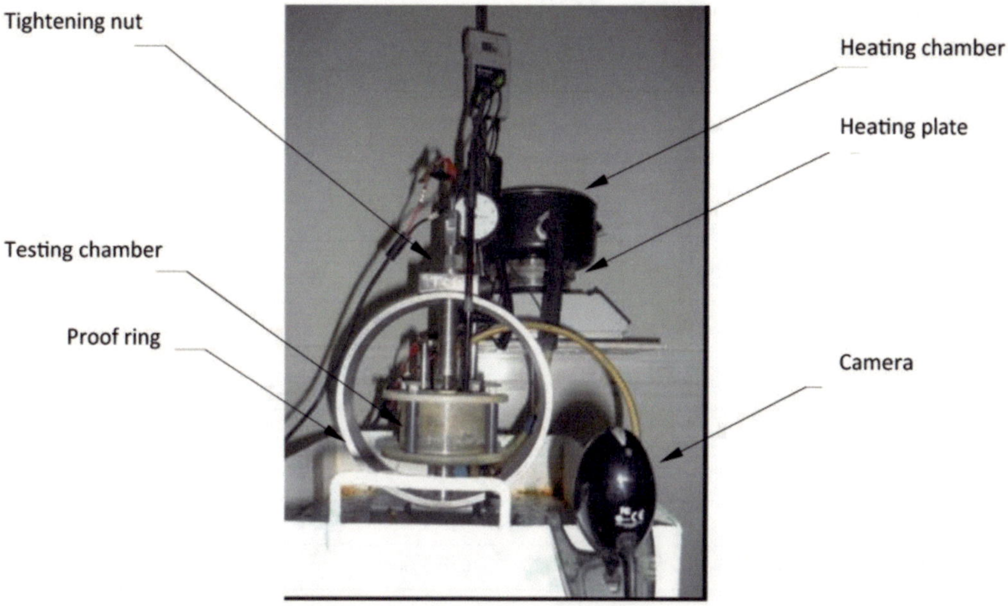

Figure 1. SCC test rig designed following the guidelines of ASTM and NACE Type A testing method [4, 5]. a) proof ring, b) SCC testing chamber, c) tightening screw and nut system, d) hot brine container, e) heating plate, f) monitoring camera [2].

A Teflon coated aluminum container, with an over flow floating valve, is used to heat the brine received from a higher level supply tank. A hot plate with a controlled power switch is used to heat the brine in the container to the required testing temperature. The heated brine is then delivered by gravity to the SCC testing chamber and hence to a disposal tank. A web monitoring camera is mounted and adjusted to record one shot each 30 min in order to detect movements of the dial indicator and hence failure of samples. Auxiliary and reference electrodes are immersed inside the testing chamber through the top Teflon cover. An ACM potentiostat (model Gill 6) is used to apply the required accelerated anodic potential during SCC testing. ACM Sequencer software is used to record the test results. An offset anodic potential with respect to the rest potential of each tested sample is normally used. The value of this accelerating anodic potential is determined from cyclic sweep and depends on the required degree of acceleration and any observed pitting potential values. During SCC testing the sample is subjected to a constant load representing a high ratio of the yield load of the tested sample. Each SCC test is stopped upon sample fracture or completion of predetermined value of testing hours, whichever comes first. Samples which are not completely separated into two pieces, by SCC tests, are subsequently forced to mechanical tensile fracture using the MTS testing machine. Fracture sections of the mechanically forced fractured samples can be examined using the scanning electron microscopy SEM. These sections can be also compared with fracture sections of fresh samples not subjected to SCC testing. The ultimate tensile loads of both fresh and mechanically forced fractured samples can be also compared.

3. Ductile Ni resist- cast irons DNI

DNI are highly alloyed class of cast irons. Their main alloying element is Nickel and its content varies between 18-22% as per relevant standards giving its austenitic microstructure and its desirable corrosion resistance properties. Other alloying elements such as chromium are present even though in lower percentages than nickel. Ni-resists come in a variety of compositions depending on their intended applications. For sea water applications which include brine circulation pumps, chemical compositions of two common grades of ductile Ni-resist in relation to the permissible range of composition as per the ASTM A439 D2 are indicated in table 1 [6].

Grade	C	Si	Mn	P	Ni	Cr	Mg	Nb	Cu
ASTM [A439 D2]	Max 3.0	1.5-3.0	0.7-1.25	Max 0.08	18-22	1.75-2.75	-	-	-
D- Material : ASTM	2.69	2.58	0.83	0.013	18.9	2.12	0	0	0
G- Material: [BS3468 S2W]	2.77	1.94	1.03	0.015	20.1	1.66	0.043	0.15	0.08

Table 1. Reported chemical compositions of the D and G-types ductile Ni-resist irons in relation to the permissible range of composition as per the ASTM A439 D2 [6]

The authors of this chapter have investigated the corrosion failure of the pressure parts of brine circulation pumps, made of DNI, in a desalination plant located on the Arabian Gulf [6, 7]. Two brands of pumps had been reported to have different lives to total failure; one lived 18 years while the other lasted only five years. The failed parts of former pumps were made out of DNI material as per ASTM A439 D2 (denoted in table 1 by D-material), whereas, those of latter pumps were made out of DNI material as per BS 3468 S2W (denoted by G-material), which has better weldability.

The material factor, as one of other possible factors that could have contributed to this different behavior, has been evaluated. Metallurgical examinations using scanning electron microscopy (SEM), image analysis, tensile tests and Vickers hardness tests were used to study the microstructure, and mechanical properties of both alloys. Electrochemical and SCC tests were performed in brine solutions to evaluate the corrosion and SCC behaviors of both alloys. The following represents a summary for the experimental work, results and conclusions of this investigation.

3.1. Experimental work

3.1.1. Image analysis and mechanical testing

Samples for all types of tests were cut from failed parts of the brine circulation pumps. Samples for metallurgical examinations were, ground, using a rotary grinder with emery paper grades up to 2400. Ground specimens were polished using a rotary polishing machine with

http://dx.doi.org/10.5772/52318

diamond paste up to 0.25 μm. Samples were then etched using 2% Nital solution (2%Nitric acid in 98% Ethanol).

Classification of graphite nodules, in both types of cast irons, in terms of average nodule diameter, number of nodules per square millimeter and average aspect ratio of nodules were determined using SEM images and Ks 300 Kontron Elektronik image analysis software. Hardness and tension tests were conducted using standard Vicker harness tester and 100 kN MTS tensile testing machine respectively. Tensile test specimens, having a gauge diameter of 12.5 mm were prepared from both types of cast iron and tension tests were conducted as per ASTM standard [8].

3.1.2. Preparation and examination of specimens with cracks induced during plant service

To permit SEM examination of the fracture surface initiated by SCC during service one specimen each from both D and G materials of approximately 2 in. x 1 in. x 1 in. and having cracks with crack front were cut out from failed pump casings. Threaded holes were prepared and special fixtures were fabricated to open the crack surfaces using an MTS machine without damaging the crack surfaces. Fig. 2a illustrates a specimen ready for crack opening. Hexa methylene tetra amine solution was used according to ASTM G1 [9] to remove as much as possible of corrosion products from fracture surface. Two other specimens, as shown in Fig. 2b, with SCC cracks were also sampled from failed pump casings and prepared for optical microscopy to examine nature of crack propagation in the matrix and other present phases.

(a) (b)

Figure 2. Two photographs of (a) G specimen prior to crack opening with the SCC appearing at top of front face and (b) G and D specimens, sampled from failed pump casings with cracks for optical microscopy [7].

3.1.3. Electrochemical testing

Hollow cylindrical test specimens having 12.4 mm outside diameter and 7.94 mm height of each type of cast iron, were machined from pieces cut from real failed pumps. Following machining, specimens were stress relieved as per the ASTM guidelines [10], chemically cleaned as per the ASTM [9] and mechanically ground using emery papers up to 600 grade. Specimens were then degreased with acetone and cleaned with fresh water prior to electrochemical testing. The corrosive environment used was brine solution (concentrated sea water of Arabian Gulf) having an average chloride concentration of 34,000 ppm. This was arranged from the desalination plant where pump failures have occurred. ACM potentiostat and software system were used for testing. The test apparatus and shape of test specimens used in electrochemical and SCC tests are shown in Fig. 3 [6]. Prepared specimens were subjected to both long term linear polarization resistance (for corrosion rate determination) and rest potential measurements. These measurements were carried out over a period of about two days at room temperature (25 ± 2°C). The above tests were directly followed by potentiodynamic sweeps to compare between the corrosion behaviors of both alloys.

(a) (b) (c)

Figure 3. Setup of electrochemical testing (a), an electrochemical test sample (b), and an SCC test sample (c)[2, 6].

3.1.4. SCC tests

The materials of both D and G materials were cut out from pump casings that failed by SCC. Locations of cuts were selected to be as near to crack area as possible. This is to ensure, to the extent possible, that microstructures of test materials are not different from that of the cracked areas. According to ASTM [11] standard A370 tensile round SCC test specimens having a small size gauge diameter of 6.25 mm were machined from the cut D and G testing materials, see Fig. 3.c. Machined specimens were subjected to stress relief heat treatment, according to ASTM A439 [9], and mechanically cleaned with emery papers to remove oxide scales resulting from the stress relief process. The specimens had then their gauge surfaces ground to 600 grit size. A Resin coating was applied at fillets and shoulders of the specimens to seal the test cell at specimen insertion holes from brine leakage. The specimen portion exposed to brine during stress corrosion testing is 20 mm of the gauge length. The threaded portions were taped to keep them clean. After the SCC testing the fracture surfaces of failed SCC test specimens were chemically cleaned using hexa methylene tetra amine solution [10] to remove as much as possible of corrosion products.

In order to accelerate the SCC testing, the specimens were subjected to following test conditions. First, air was allowed to contact the brine stored in an overhead tank. Second, continuous supplies of brine ensure that the brine in the test cell stays fresh and rich in chlorides hence maintaining its corrosiveness. Third, the brine solution temperature was raised to around 55 °C. From Miyasaka's work it was found that this temperature is high enough to significantly accelerate SCC of DNI [12]. Fourth, specimens were all anodically polarized by 100 mV with respect to their free corrosion potential. Finally, all specimens were highly stressed between 73% and 102% of their respective 0.2% offset yield strengths. A total of eight specimens; four from each alloy were SCC tested. Table 2 shows the details of the stress levels of the tested samples. The applied stresses were chosen as follows. Two specimens from each of the G and D materials were tested at around 220 MPa. This is approximately 100% and 86% of the yield stresses of the G and D material respectively. For unbiased comparison, two G specimens were also tested at around 86% of their yield stress. Two other D specimens were tested at stress levels around the 86% and 93% of the yield stress of the G material (73.3% and 79.2% of the yield stress of the D material) for comparison. Stressing the specimens was carried out, using the proof ring described earlier in this chapter, after inserting the specimen in the test cell.

Sample	G1	G2	G3	G4	D1	D2	D3	D4
Stress, MPa	225	216.5	190.4	190.4	224.5	216.5	190.6	205.9
%, of yield*	102	98.4	86.5	86.5	86.4	83.4	73.3	79.2

*Percentage of 0.2% offset yield stress of each alloy.

Table 2. Details of the stress levels of the tested samples [7].

Once the specimen was stressed various connections were made. The heater was set at 55 °C using a heater element thermostat and a PC-connected camera was then hooked around the test cell using rubber bands. The camera software was set so as to take photos at intervals of 15 min. The specimen was then polarized using ACM potentiostat and software. After admitting the brine solution and reaching the test temperature at (55 ± 3 °C), the camera was activated, the potentiostat was run and the time to full fracture was recorded visually with the camera.

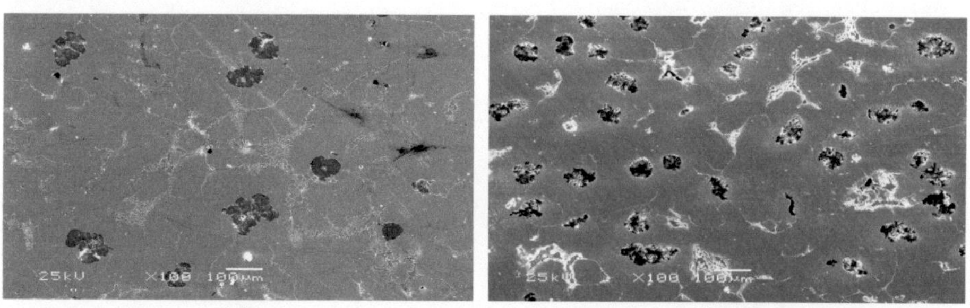

(a) (b)

Figure 4. SEM micrographs showing the difference between the microstructures of the D-type (a) and the G-type (b) of the Ni-Resist austenitic cast irons [6].

3.2. Results and discussion

Fig. 4 shows SEM micrographs, illustrating the difference between the microstructures of the D-type (a) and the G-type (b) of the Ni-resist ductile irons [6]. Table 3 shows the image analysis results for both types of cast iron. These results show that the number of graphite nodules per square millimeter for the D-type cast iron is almost half the number of that for the G-type. However, the average nodule diameter of the D-type is greater than that for the G-type.

Table 3 also shows that the graphite nodules of the D-type are more circular in cross section than the nodules of the G-type cast iron. This is illustrated by the higher average aspect ratio of the D-type nodules as compared to the average aspect ratio of the G-type nodules.

	D-type	G-type
Field area (mm2)	4.505	4.505
Number of nodules	113	227
Number of nodules / mm2	25.08	50.39
Average nodule diameter (μm)	43.67	31.54
Average aspect ratio	0.717	0.645
Percentage area of graphite to the total field area	3.8 %	2.09 %

Table 3. Image analysis of the D and G-Types of the Ni-Resist austenitic cast irons [6].

The SEM micrographs of Fig. 4 also show that carbides are more uniformly distributed within the microstructure of the D-type cast iron. This can explain the relatively higher Vickers hardness number and tensile strength of this type of cast iron. Average Vickers hardness values, HV5, of 220 and 200 have been measured for the D and G-types, respectively. Fig. 5 shows that the D type has a higher 0.2% offset yield strength of 260 MPa as compared to the 220 MPa of the G –type. The modulus of elasticity of both materials is about 131 GPa which is on the upper side of the range reported in the standards [13]. The EDX chemical analysis within the field area of these micrographs shows that Cr is basically existing in carbides rather than being free in the matrix.

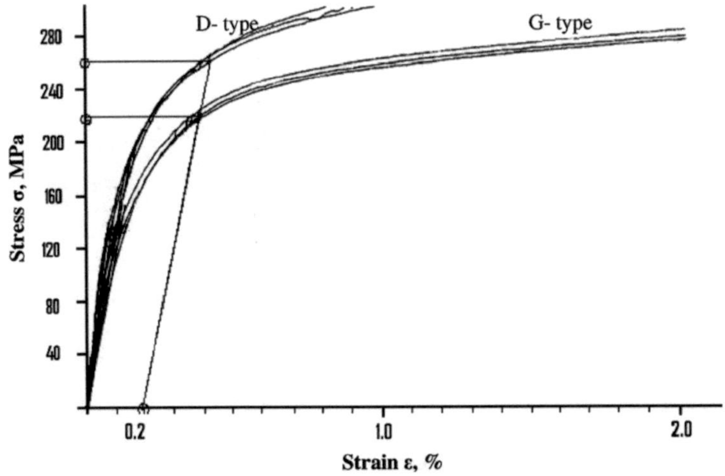

Figure 5. Stress–strain plots for the two types of nickel-resist cast irons [6].

Results of corrosion rates and rest potential measurements are shown in Fig. 6. The corrosion rates stabilized between at 0.2–0.25 mmpy and the potentials ranged between -450 mV and -500 mV, all with respect to Ag/AgCl reference electrode. Similar results have been reported in literature [14]. Fig. 7 shows the Tafel plots for the two cast iron materials. Both types showed similar behavior in shape of curves even though the rest potentials varied from -500 mV to -650 mV without any distinctive pattern for either type of materials. Severe corrosion process took place at potentials greater than 100 mV (Ag/AgCl RE) anodic to the rest potential. This was accompanied by blackish thin corrosion layer and rigorous bubbling at the surface of the cylindrical counter electrode (made of duplex stainless steel). Similar results in synthetic sea water environment (3% NaCl) solution at 25 °C, have been reported [15].

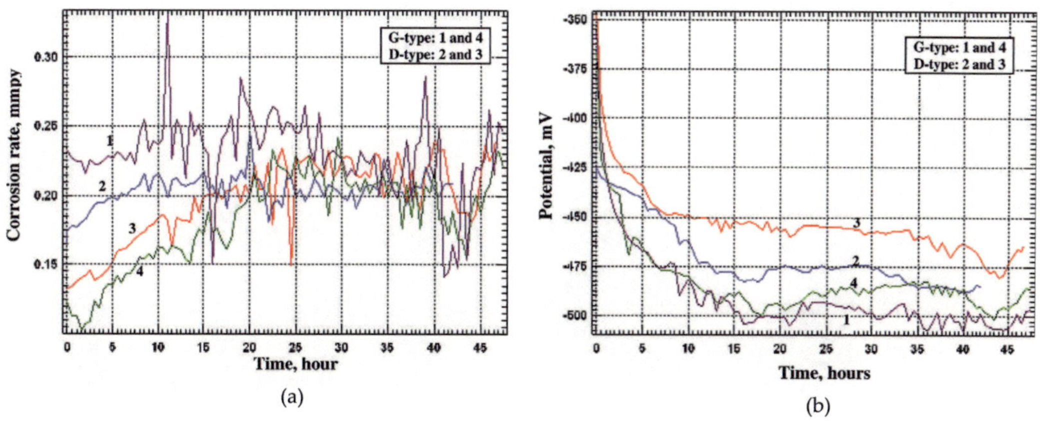

Figure 6. Corrosion rates (a) and potentials (b) of four specimens (two from each D and G materials). Potentials are measured versus Ag/AgCl reference electrode [6]

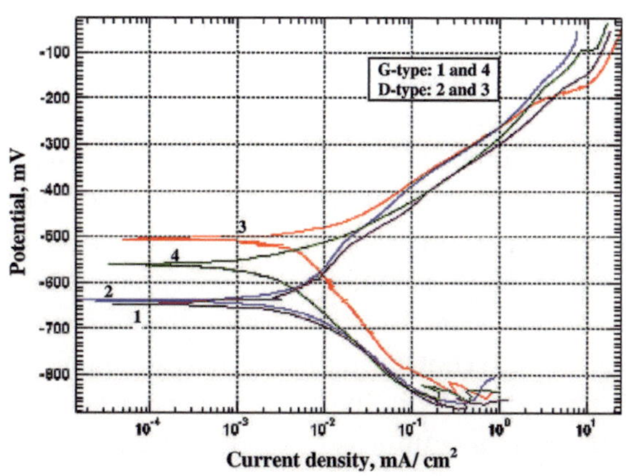

Figure 7. Tafel plots of four specimens (two from each D and G-type materials). All potentials are measured versus Ag/AgCl reference electrode [6].

Thus, electrochemical corrosion tests, in brine solution at room temperature, have shown similar corrosion behavior, in terms of corrosion rates, potential and polarization. To compare between the combined effects of strength and corrosion resistance of both alloys, SCC tests were performed.

The two specimens, sampled from failed pump casings, for examination of the crack surfaces were opened till fracture using special holders prepared for this purpose. Figure 8 shows two photographs for the fracture surfaces of the two materials. They show corroded and mechanically fractured areas. Figure 9 shows photographs for the service induced SCC of the two materials. All cracks pass through the matrix without preference to phases. Figure 10 shows SEM micrographs of fracture surfaces of the same specimens shown in figure 8. They indicate two distinctive fracture surfaces. The first is the mechanical fracture surface caused by loading using the MTS tensile testing machine and the second is the SCC fracture surfaces developed during pump service. The second surfaces are similar to those reported in literature [16].

(a) (b)

Figure 8. Two photographs showing (a) G material as fractured prior to cleaning and (b) D material after chemical cleaning [7].

The life times to full fracture of various tested specimens are indicated in Table 4. As can be seen from the table, all G specimens fractured during the tests. Out of the 4D specimens only specimen D1 fractured. These results tend to agree with times to failures reported in actual plant service with D material outperforming G material (G failed in 5 years whereas D failed in around 18 years). Fig. 11 shows SEM micrographs of fracture surfaces of SCC tested D and G materials. These micrographs of fractured surfaces clearly indicate a fracture pattern of two different surface morphologies. While much of the fracture surface has dimpled non-flat areas characteristic of purely mechanical fracture, there are flat areas extending from edges of the specimen. They also contain sudden vertical steps and transverse cracks do exist in the flat areas in many instances. They also are characteristic of transgranular SCC agreeing with those available in literature as reported by Kauczor [16]. This fracture pattern matches with the fracture pattern obtained and presented in Fig. 9 for the samples collected from the casings of the service failed pumps. This emphasizes that the cause of failure of the casings of the failed pump is stress corrosion cracking.

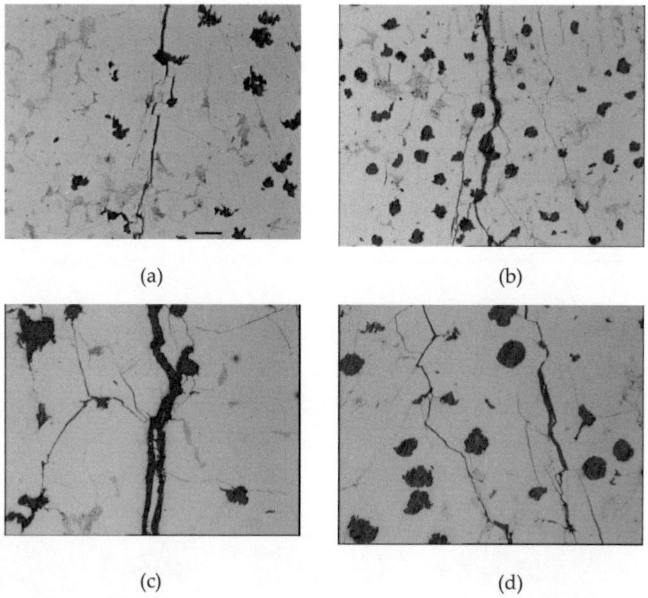

Figure 9. Service induced SCC cracks of D material at (a) 100X and (c) 200X and those of G material (b) 100X and (d) 200X. Cracks propagate through matrix without preference to phases [7].

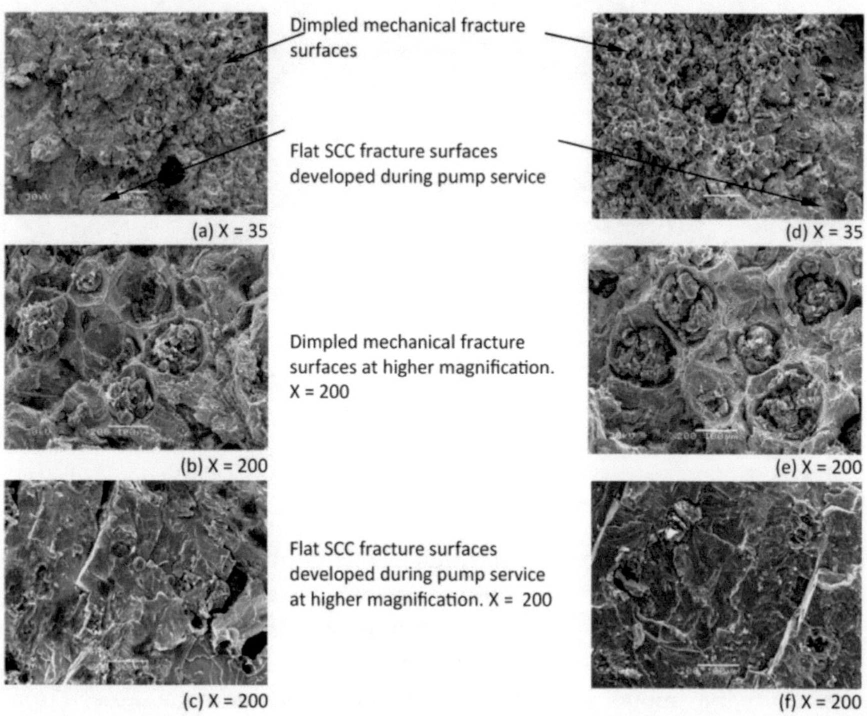

Figure 10. SEM micrographs of fracture sections of D material (a, b, and c) and G materials (d, e, and f). SCC fracture surfaces were produced during pump service, whereas, mechanical fracture surfaces were produced due to specimen forced fracture using MTS testing machine [7].

Sample	G_1	G_2	G_3	G_4	D_1	D_2	D_3	D_4
Applied stress, MPa	225	216.5	190.4	190.4	224.5	216.5	190.6	205.9
%, of 0.2% offset yield*	102.3	98.4	86.7	86.5	86.4	83.4	73.3	79.2
Time to failure, hours	37.6	86.5	100.5	72.5	167.5	184.8- TS**	254- TS	209- TS
Remarks***	FF	FF	FF	FF	FF	NF	NF	NF

*0.2% offset yield of G = 220 MPa. 0.2 offset yield of D = 260 MPa [6] ** TS = Test stopped without fracture *** FF = Full fracture, and NF = No fracture.

Table 4. Time to SCC failure of tested specimens of. G and D materials [7].

The above results suggest that as other factors are neutralized the material factor has a significant role in the reported contrasting performance of DNI with respect to resistance to SCC.

(a) X = 70

Photograph (a) shows SCC fracture surface at the upper right portion and mechanical fracture surface at the lower left portion.
Photograph (b) is the elected area of (a) at X = 200.

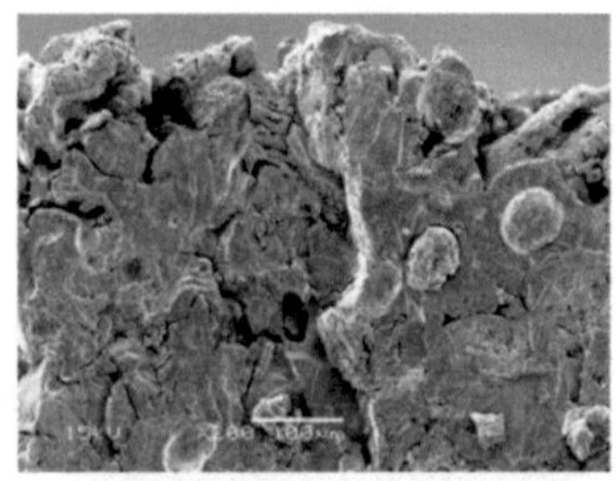

(b) X = 200

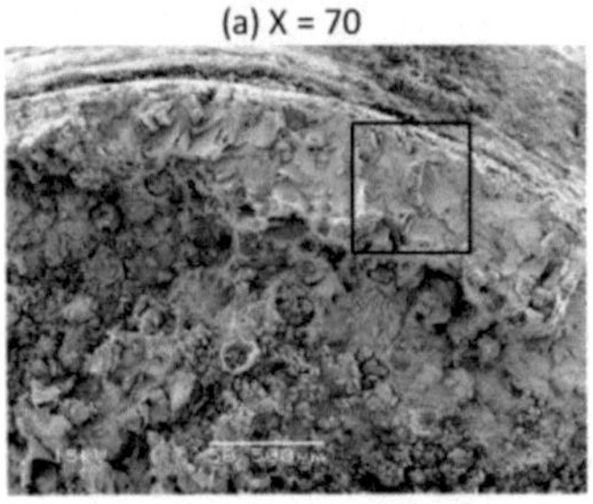

(c) X = 50

Photograph (c) shows SCC fracture surface at the upper portion and mechanical fracture surface at the lower portion.
Photograph (d) is the elected area of (c) at X = 250.

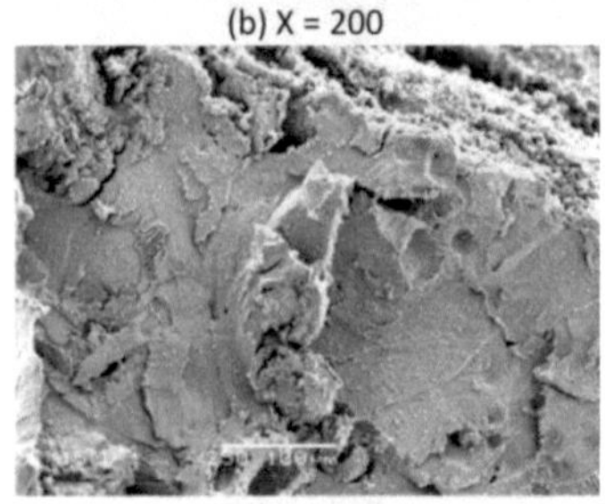

(d) X = 250

Figure 11. SEM micrographs of G material, G3 specimen, (a), (b) and D material, D1 specimen, (c), (d) showing fracture surface pattern similar to that of failed pump casing shown in figure 9 [7].

In addition to SCC the specimens were simultaneously subjected to uniform corrosion under anodic polarization applied to accelerate the SCC. The effect of uniform corrosion on specimens' final state of stress was examined in accordance with ASTM G49 [5]. This has been done by calculating the clean cross sectional area of each specimen, after SCC testing, and consequently the amount of increase in the applied stress. It can be seen from Table 5 that the average percent in diameter reduction in D material is relatively higher than that observed in G material. This can be attributed to the longer periods of testing of the D material. The consequent average stress rise in D material is comparable with that of G material.

http://dx.doi.org/10.5772/52318

However, such increased stresses are still much below the ultimate stresses of both materials. This indicates that the obtained SCC testing results are not biased by area reduction due to uniform corrosion.

The above contrasting behavior in SCC resistance of these two materials can be explained in view of the variation in their yield stress. The 0.2% offset yield stress of D material is higher than that of G material by approximately 40 MPa [6]. According to Miyasaka and Ogure [12] the log of time to failure by SCC is inversely proportional to the applied stress. Even though the ultimate stresses of both materials are approximately equal, the yield stress would practically have a more pronounced effect on SCC resistance. This view point is supported by the fact that SCC takes place at lower stresses than the yield stress as seen in Table 4 and reported by Miyasaka and Ogure [12]. As can be seen from Fig. 5, under a stress value of say 260 MPa, which is the yield stress of D material, G material would be subject to a strain value of around 1% whereas D material would be strained only to a value of 0.33%. This difference in matrix stretching would certainly make G material more prone to SCC as compared to D material. Another possible reason for this difference in performance is the characteristics of graphite nodules in each material.

Specimen	Load, N	Initial Dia., mm	Initial Stress, MPa	Final Dia., mm	Reduction in Dia., %	Stress Rise*, MPa	Final Stress as % of Ult. Stress
G_1	6836	6.22	225	6.11	1.77	8.2	65.7
G_2	6452	6.16	216.5	5.86	4.87	22.7	67.4
G3	5723	6.18	190.8	6.00	2.91	11.6	57.0
G4	5856	6.26	190.3	6.18	1.28	5.0	55.0
D1	6757	6.19	224.5	6.09	1.62	7.4	66.9
D2	6190	6.19	216.5	5.80	6.30	17.8	67.5
D3	5940	6.30	190.6	6.15	2.38	9.4	57.6
D4	6379	6.28	205.9	6.09	3.03	13.0	63.1

*Excluding stress rise due to crack effect on section reduction.

Table 5. Effect of specimen reduction in cross sectional area, due to uniform corrosion, on final state of stress [7].

As can be seen from Table 3, the nodules of D material are bigger and fewer in number than G material. The former has an average diameter that is around 40% larger than the latter's average diameter. Also number of D material nodules are half that of G nodules in the same size of field area. As the SCC is a surface phenomenon taking place at the material surface in contact with the corrosive environment, the size of graphite nodules and their number may be significant. The nodules are non-load bearing and incoherent phase in the iron as clearly shown in SEM micrographs of Fig. 10 which illustrate voids left by nodules and gaps between the matrix and periphery of exposed nodules. Raman [17] has studied the caustic SCC of ductile iron. He found that "where crack encountered graphite nodules, further propagation involved decohesion in the nodule-matrix interface". As such, surface nodules can be considered as micro-cracks or notches. From a fracture mechanics view point the smaller the diameter of these natural notches the more is the stress concentration at these points. The nodule count may also have contributed to the different behavior in resistance to SCC. The higher the number of nodules at the exposed surfaces, as in G material, the higher is the possibility of crack initiation and propagation. This is again supported by Raman's findings [17] and makes the time to failure by SCC of G material shorter compared to D material. It was [14] indicated that ''assigning of degrees of susceptibility (to SCC) is of questionable merit.'' To the contrary to this statement the results in this study indicate that ranking of Ni-resists with respect to SCC resistance is viable. This is also in agreement with what Miyasaka and Ogure [12] had reported. Further, the results clearly indicate that the relevant standards [10, 18] for ductile Ni-resists do not provide the required protection against SCC in marine service even after subjecting the cast materials to suitable stress relief heat treatments, again in contrary to what was reported [14] above. For better field performance the standards need modifications based on further studies with regards to mechanical and microstructure properties. This might include carbide characteristics and nodule features so as to arrive at an optimized microstructure leading to best resistance of DNI to SCC in marine environment. Such modifications would necessitate more stringent quality control and assurance procedures in manufacturing facilities. Meanwhile, super duplex stainless steels have found wider use in marine service [19] in recent years and many brine and sea water pumps got their failed DNI casings replaced with such superior materials.

4. Stainless steels

Two types of stainless steel are suggested and recommended to substitute Ni resist iron in manufacturing pump casings [3]. These are austenitic stainless steel UNS S31603 and super duplex stainless steel UNS S32750 [20]. In addition to cast stainless steels, original pump manufacturers sometimes use welded construction of wrought stainless steels to build other related components such as column pipes and discharge elbow piece [21]. In fact the desalination plant, in which failures of brine circulating pumps have been reported, has used welded stainless steel S316 material as replacement for failed Ni-resist components [21]. Thus the study of SCC of wrought stainless steels is significant since the long term performance of these materials is still to be seen. In this part, the mechanical, metallurgical, electrochemical and SCC properties of the above mentioned two types of wrought stainless steels are presented through experimental investigation.

4.1. Experimental work

4.1.1. Material preparation and tensile tests

Two strips of hot rolled plates of UNS S31603 and UNS S32750 having a thickness of 12.7 mm were cut into samples having dimensions of 250 x 50 x 12.7 mm. Chemical analysis, by weight percent of elements in each type of steel is shown in Table 6 [22, 23]. Tensile test samples for each type of stainless steel were prepared as per the ASTM standard A 370-07

http://dx.doi.org/10.5772/52318

[24].Machining of all samples was carried out using a machining coolant to avoid samples overheat.

Material	C%	Mn%	P%	S%	Si%	Cu%	Ni%	Cr%	Mo%	N%	Co%
UNS S31603	0.025	1.360	0.029	0.003	0.268	0.468	10.056	16.804	2.176	0.051	0.213
UNS S32750	0.017	0.893	0.031	0.0004	0.370	0.126	6.651	24.681	3.755	0.280	-

* The balance each composition is iron.

Table 6. Chemical composition of the two as received types of stainless steel

4.1.2. Metallographic and hardness tests

Samples from the as received material of both austenitic UNS S31603 and super duplex UNS S32750 were cut into thin sections using a thin sectioning cutter. A cutting coolant was used during cutting to avoid overheating. Thin sections were then mounted in phenol moulds to be ready for grinding and polishing. Grinding emery papers having grids of 240, 400, 600, 1000, and 2400 were used. Polishing was performed in two stages using 6 μm, and 1 μm diamond pasts. Austenitic stainless steel samples were electrolytically etched using 10% oxalic acid at 3 V, whereas, super duplex samples were etched using an electrolyte of 20% NaOH and 100 ml of distilled water at 3 V for 20 s. Samples were then examined using an optical microscope and a digital image camera was used to capture microstructures of both steels. For hardness tests, a load of 200 g was applied on different locations of the microstructure for each tested sample by using Vickers micro hardness testing device. For each type of steel, an average of five different readings has been calculated.

4.1.3. Electrochemical tests

4.1.3.1. Sample preparation

Three sets of polarization test samples from the as received austenitic and super duplex stainless steels were cut and machined to dimension of 70 x 10 x 5 mm. Each set, consisting of one super duplex and three austenitic stainless steel samples, was connected to insulated electrically conducting wires. The assembly was then molded in an epoxy mixture consisting of a resin and a hardener to ensure complete electrical insulation among samples. Extra care was taken during molding to avoid having air gaps between stainless steel samples and epoxy and hence avoid possible crevice corrosion. Molded samples were manually ground with 100, 200, 400, 600, and 1000 emery papers, degreased using 5% caustic soda solution and rinsed in fresh water. The electrical wires were passed through water seal PVC tubes and fittings for perfect insulation during immersion in brine solution. Ends of wires were identified and labeled as, austenitic stainless steel working electrode, super duplex stainless steel working electrode, and two additional austenitic stainless steel samples to serve as reference and auxiliary electrodes as shown in Fig. 12 [2].

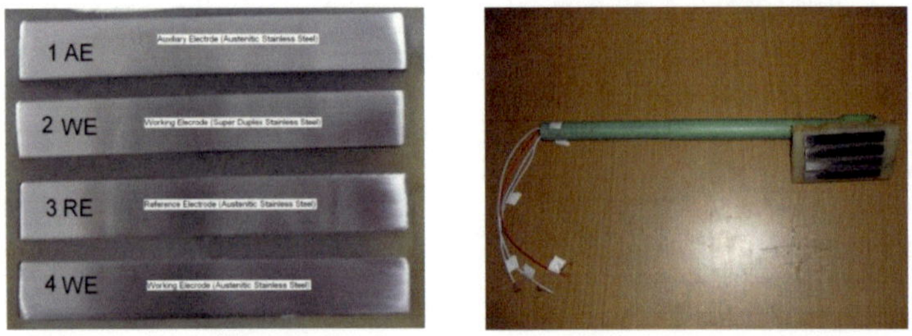

(a) (b)

Figure 12. a) Identification of electrodes in one sample set. (b) Molded samples in epoxy and copper wire passing through water seal PVC tube [2].

4.1.3.2. Long term potential measurements

An ACM potentiostat Gill 8 connected to a computer was used to perform the electrochemical tests, whereas, Sequencer software was used to control and record the test results. Samples were immersed in Pyrex container filled with a temperature controlled brine solution. The brine solution, which is a concentrated sea water of Arabian Gulf having an average chloride concentration of 34,000 ppm, was arranged from the desalination plant where pump casing failures have occurred. Open circuit potentials of the as received austenitic and super duplex stainless steel samples were measured using long term potential measurements. Samples were immersed in brine at 60 °C and pH of 8.31 for 1 h before running the test to take potential measurements for duration of 1 day.

4.1.3.3. Cyclic sweep

Cyclic sweep testing was performed under the above mentioned conditions on stainless steel samples of both types. For austenitic stainless steel, the start potential was set to -250 mV and the reverse potential to +750 mV with reference to its open circuit potential. For super duplex stainless steel, the start potential was set to -250 mV and the reverse potential was set to +1000 mV with reference to its open circuit potential. The sweep rate was 30 mV/min.

4.1.4. SCC tests

4.1.4.1. Sample preparation

Three SCC samples from each type of stainless steel were machined from the as received rolled plates to conform with the NACE type "A" SCC test method [4] and ASTM standard G49 [5]. Machined SCC test samples have gauge diameter and length of 6.24 mm and 40 mm respectively. All samples were machined to have the same dimensions. Machining of samples was carried out using a coolant to avoid sample overheating. Samples were manually ground with 100, 200, 400, 600, and 1000 emery papers, degreased using acetone solution, and rinsed in fresh water [4, 5, 24].

4.1.4.2. SCC testing

The testing rig and method described earlier in this chapter in section1 (SCC testing rig and method) was used to conduct SCC tests. An ACM potentiostat model Gill 6 was used to apply the required accelerated anodic potential during SCC testing. ACM Sequencer software was used to record the test results. An offset anodic potential of +400 mV with respect to the rest potential of each of the as received austenitic and super duplex stainless steels was used [25]. The value of this accelerating anodic potential was determined from cyclic sweep based on the pitting potential values observed for the austenitic stainless steel samples. During SCC testing, both types of stainless steel samples were subjected to a constant load of 8403 N representing 95% of the yield load of the as received austenitic stainless steel and 43% of the yield load of the as received super duplex stainless steel. Each SCC test was stopped upon sample fracture or completion of 335 testing hours (14 days), whichever comes first. Samples which were not completely separated into two pieces, by SCC tests, were subsequently forced to mechanical tensile fracture using MTS testing machine. Fracture sections of the mechanically forced fractured samples were examined using SEM. These sections were also compared with fracture sections of fresh samples not subjected to SCC testing. The ultimate tensile loads of both fresh and mechanically forced fractured samples were also compared.

4.2. Results and discussion

Table 7 shows the average mechanical testing results of both types of steels. Austenitic steel enjoys better ductility on the expense of its yield and tensile strengths as compared to super duplex steel. The table illustrates considerable differences in the yield and ultimate strengths of the two types. Results of hardness testing have also shown a noticeable difference between both types. Austenitic steel was found to have an average hardness of HV 202.6 as compared to HV 265 for the super duplex steel.

	Yield Strength N/mm2	Ultimate Tensile Strength N/mm2	Elongation %	Reduction of Area %
Austenitic UNS S31603	284	597	52	73
Super duplex UNS S32750	608	852	35	68

Table 7. Average mechanical testing data of the as received two types of stainless steel [2].

Fig. 13 shows micrographs for the austenitic steel microstructure at two magnifications. The micrographs illustrate grains of two phases. The austenite which is the majority phase appears in the micrograph as the light phase and ferrite which is the minority phase appears as the dark phase. Elongated ferrite grains indicate the rolling direction. Fig. 14 illustrates micrographs for the as received super duplex stainless steel UNS S32750, showing that ferrite is the majority phase and austenite is the minority phase.

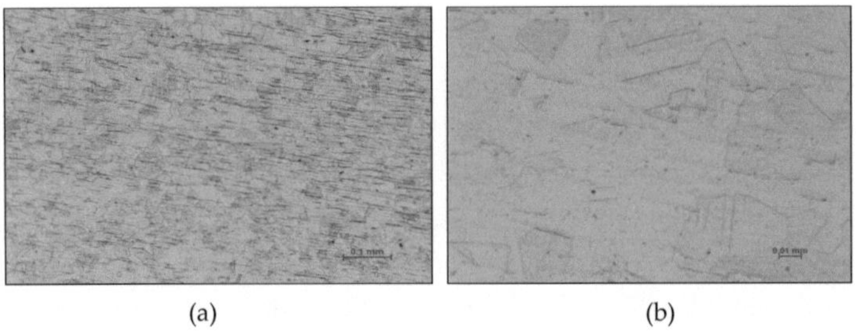

Figure 13. Microscopic images for the austenitic stainless steel UNS S31603 (a) at 100X and (b) at 500X [2].

Fig. 15 shows the open circuit potential graphs for the as received austenitic and super duplex stainless steels. All potentials are measured against austenitic stainless steel as the reference electrode. The average values of the open circuit potentials recorded for the last 4 h of testing for both types are shown in Table 8. These values indicate a relatively lower corrosion tendency for super duplex steel. This behavior is schematically illustrated in Fig. 16 using E-Log i diagram.

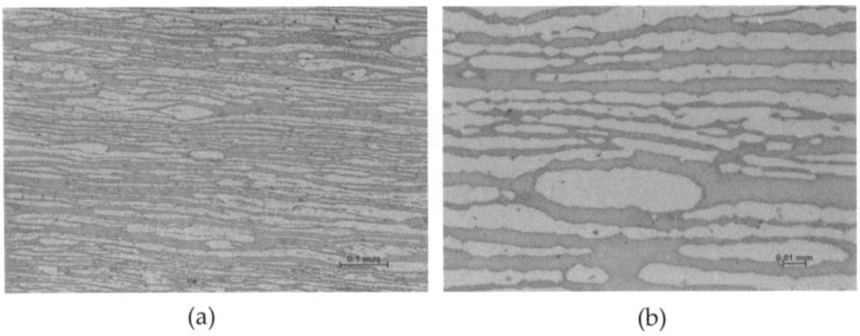

Figure 14. Microscopic images for the as received super duplex stainless steel UNS S32750, (a) at 100X and (b) at 500X [2]

Fig. 16 shows that when a relatively higher cathodic potential is measured on super duplex stainless steel a correspondent lower current density is expected. On the other hand, higher anodic potential corresponds to higher current densities on the surface of the austenitic stainless steel. The lower corrosion tendency of the super duplex stainless steel is clearly attributed to the relatively higher chromium levels which help form a more stable and a stronger passivity on its surface under these specific test conditions. The enhanced passivity of the super duplex was also confirmed visually by having no signs of pitting and by electrochemical cyclic sweeps.

Material	Austenitic stainless steel UNS S 31603	Super duplex stainless steel UNS S 32750
Average open circuit potential, mV	7.13	-24.01

Table 8. Average open circuit potentials for austenitic and super duplex steels during the last four hours of testing[2]

http://dx.doi.org/10.5772/52318

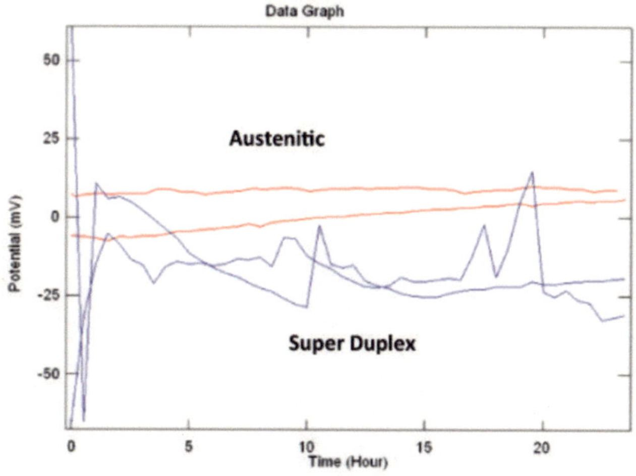

Figure 15. Open circuit potential against time for the as received austenitic and super duplex stainless steels. All potentials are measured against austenitic stainless steel reference electrode [2].

Cyclic sweep test plots are shown in Fig. 17 for austenitic and super duplex stainless steels. Results of these tests indicate that localized break down in the passivity in the form of a clear pitting takes place for austenitic steel at any potential between 275 and 398 mV measured versus austenitic reference electrode. The pitting behavior on the surface of the austenitic stainless steel was also confirmed by visual inspection of tested samples.

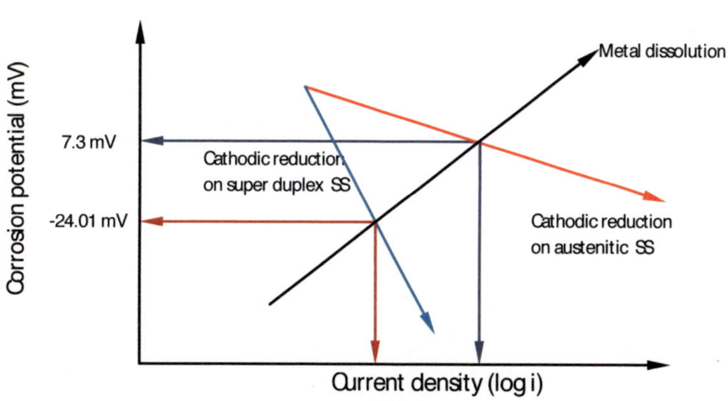

Figure 16. Schematic for (E- log i) diagram illustrating higher corrosion tendency for austenitic stainless steel versus relatively lower corrosion tendency on super duplex stainless steel [2].

Fig. 18 depicts photographs showing pitting of the austenitic sample. The super duplex specimens showed no signs of pitting after cyclic sweep testing. Similar results have been reported in literature [26, 27].

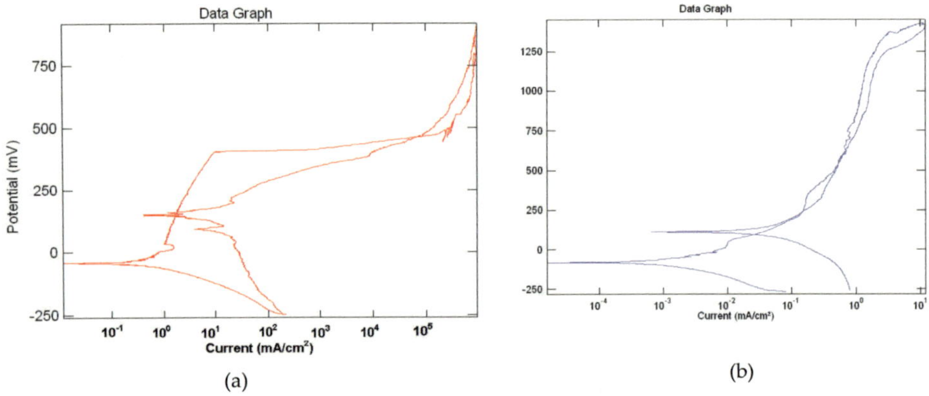

Figure 17. Cyclic sweep potential plots against current density for the as received (a) austenitic and (b) super duplex stainless steels. All measured potentials are measured versus austenitic stainless steel reference electrode[2].

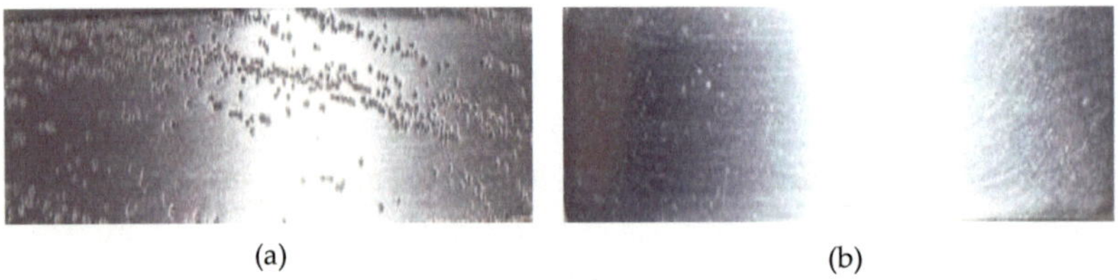

Figure 18. Pitting on austenitic sample (a) as compared by no pitting on super duplex sample (b) after one run of cyclic sweep testing [2].

Table 9 shows the recorded time to failure of both types of stainless steel due to SCC testing. The table shows that the first and third samples of austenitic stainless steel failed, after 160.29 h and 119.56 h respectively. Both samples were failed without complete separation into two pieces. The second sample failed after considerable less time of 76.38 h with complete sample separation.

Material	Time to failure, hours Austenitic steel UNS S 31603	Time to failure, hours Super duplex steel UNS S 32750
Sample 1	160.29[1]	No failure[3]
Sample 2	76.38[2]	No failure[3]
Sample 3	119.56[1]	No failure[3]

[1]Failure without complete sample separated into two pieces. [2]Failure with complete sample separation into two pieces. [3]SCC test stopped after 335 hours.

Table 9. Time to failure for austenitic and super duplex stainless steels due to SCC testing [2].

The maximum recorded subsequent tensile load required to bring the first sample to complete separation was found to be 2911.81 N versus 18,640 N; the average maximum tensile load for fresh samples (15.62% remaining strength). The remaining strength of the third sample was calculated and found to be 43.45%. In all three samples of austenitic stainless steel, it was noticed that, during SCC testing, pits were developed on their surfaces. This was confirmed by further visual examinations by the end of each test. These examinations revealed also many small cracks and some of them were joined together to form major ones.

Table 9 illustrates also that all super duplex three samples didn't fail and SCC tests were stopped after 335 h of testing. Visual observations showed, however, that under the reported test conditions and after 335 h of SCC testing, two super duplex samples had completely clean surfaces while the third sample had a single shallow pit on its surface. Subsequent forced tensile fracture of all samples, using the MTS machine revealed no strength losses due to SCC testing.

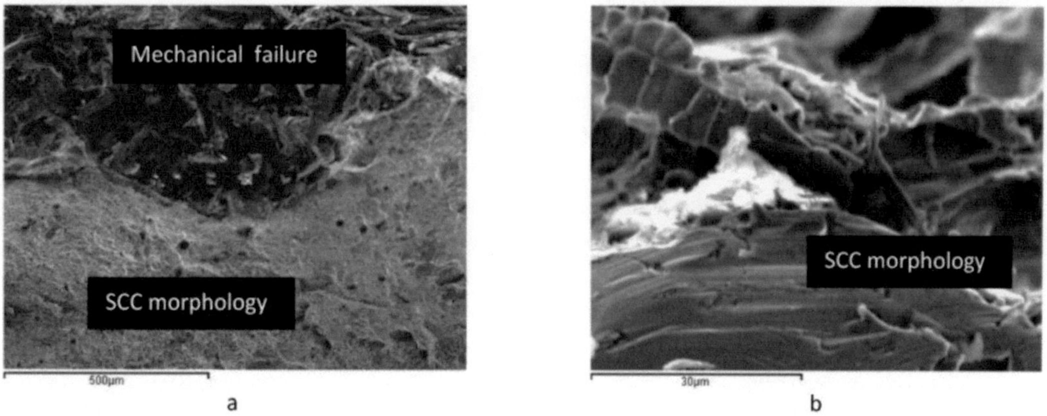

Figure 19. Fracture section of failed austenitic stainless steel sample due to SCC at X 200 (a) and X 3500(b) [2].

Fig. 19 shows fracture section micrographs of failed austenitic stainless steel sample due to SCC at two magnifications. In Fig. 19a, two different morphologies can be identified for the fracture section; one associated to the progress of cracking due to stress corrosion and the other one corresponds to fast mechanical fracture as the section is reduced due to SCC. In Fig. 19b, step like topography together with SCC facets which are analogous to cleavage facets are shown [28]. Fig. 20 shows fracture sections of super duplex stainless steel samples before and after SCC testing. Both micrographs illustrate typical ductile fracture characteristics where dimpled fracture can be identified. Comparing the micrographs of Fig. 20 reveals that their fracture surfaces are identical and that there is no sign for SCC failure.

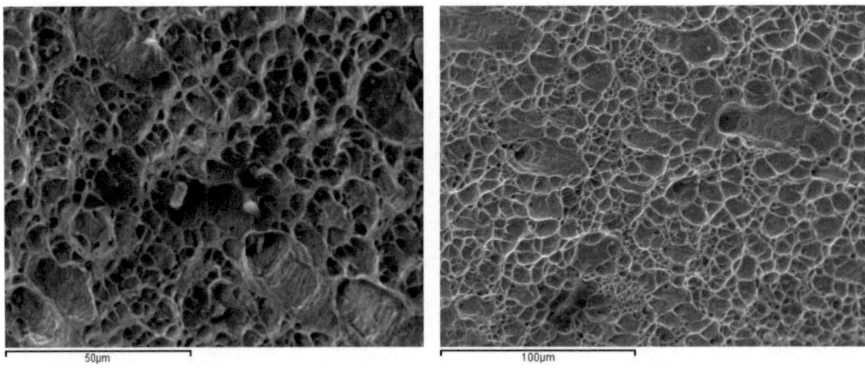

(a) (b)

Figure 20. Fracture sections of super duplex stainless steel samples before SCC testing, X 2000 (a) and after 335 hours of SCC testing, X 1000 (b). Both samples were forced to mechanical failure using the MTS testing machine [2].

5. Corrosion inhibition of austenitic stainless steel

The above results demonstrate the superiority of the SCC resistance of the super duplex stainless steel UNS S32750 over the austenitic stainless steel UNS S31603 in hot brine environment. However, due to the remarkable difference in price between these two types of stainless steels, and in case of using ASS in some pump casing metallurgies, an attempt is made to improve the immunity of the cheaper austenitic stainless steel through using chemical treatment via one proven performance corrosion inhibitor [29]. The following represents a study on the effect of using a passivating type commercially available Molybdate corrosion inhibitor on the corrosion resistance and SCC of austenitic stainless steel UNS S31603 in hot brine environment. Sodium Molybdate was selected to inhibit this type of stainless steel as it is effective at relatively low concentrations, environmentally safe, non toxic and known to passivate pits and crevices from corrosion [30]. Several works have been reported indicating the use of Molybdate solutions to inhibit different types of austenitic stainless steels in different environments [31, 32].

5.1. Experimental work

5.1.1. Sample preparation

For each of the following electrochemical tests, given in the next subsections, two sets of polarization test samples from the as received austenitic stainless steel UNS S31603, given in table 6, were prepared. One set was used as a control set for testing the electrochemical properties of the stainless steel in regular hot brine without inhibitor. The other set was used to get the same properties when the stainless steel samples were subjected to the same brine environment in the presence of a commercial type passivating Molybdate corrosion inhibitor. The treat rate of the corrosion inhibitor was 350 ppm in all treated brine solutions.

Each set of the prepared test samples consisted of four austenitic stainless steel samples each having dimensions of 70X10X5 mm and connected to insulated copper wires. Each set was prepared as described earlier in section 4.1.3 and Fig. 12. Ends of copper wires were identified and labeled as, two austenitic stainless steel working electrodes, and two austenitic stainless steel samples as reference and auxiliary electrodes. Molded samples were manually ground with 100, 200, 400, 600, and 1000 emery papers, degreased using 5% caustic soda solution and rinsed in fresh water.

5.1.2. Cyclic sweep tests

An ACM potentiostat Gill 6 connected to a computer was used to perform cyclic sweep tests, whereas, Sequencer software was used to control and record the test results. Samples were immersed in Pyrex container filled with a temperature controlled regular (uninhibited) or inhibited brine solution.The regular brine solution, which is a concentrated sea water of Arabian Gulf having an average chloride concentration of 34,000 ppm, was arranged from the desalination plant where relevant pump failures have occurred. Samples of one set were immersed in regular brine solution at 55 °C and pH of 8.31 for 24 hours before running the tests. Similar tests were performed on samples of the second set which were immersed in inhibited brine solution under the same testing conditions. The start potential was set to -250 mV and the reverse potential to +750 mV with reference to the corresponding sample open circuit potential. The sweep rate was 30 mV/ min [29].

5.1.3. SCC tests

Six SCC samples were machined from the as received rolled plates to conform with the NACE type "A" SCC test method [4] and ASTM standard G49 [5]. A machined SCC test sample has a gauge diameter and length of 6.24mm and 40 mm respectively and has already been shown in Fig. 3.c. All samples were machined to have the same dimensions. Machining of samples was carried out using a coolant to avoid sample overheating. Samples were manually ground with 100, 200, 400, 600, and 1000 emery papers, degreased using acetone solution, and rinsed in fresh water [4, 5, 24]. SCC tests were performed using the test rig described earlier in this chapter in section 2. An offset anodic potential of +400 mV with respect to the rest potential of each as received austenitic stainless steel sample was used. The value of this accelerating anodic potential was determined from cyclic sweep tests based on the pitting potential values observed for austenitic stainless steel without inhabitation. Austenitic stainless steel samples were subjected to a constant load representing 95% of the yield load of the as received austenitic stainless steel material. Each SCC test was stopped upon sample fracture or completion of 335 testing hours (14 days), whichever occurred first [29].

5.2. Results and discussion

Potentiodynamic scans (cyclic sweeps) are used to stress the metal in short laboratory periods and illustrate the performance. Cyclic sweep test plots from these experiments are shown in Fig. 21 for austenitic stainless steels without inhibition and with inhibition respectively. Results of these tests indicated that after 24 hours of immersion the open circuit potential of the austenitic stainless steel has been slightly shifted (60-70 mV) in the noble direction due to the change in the metal passivity resulting from the chemical treatment. The plots show also that, without inhibition, pitting takes place for austenitic steel at an average potential of 337 mV measured versus austenitic reference electrode, while an average pitting potential of 318 mV with inhibition was observed. Pitting was observed on all austenitic samples following cyclic sweep tests. Pits are believed and known to be the initiation sites for any possible SCC behavior that the material could undergo under specific combination of environment and stress conditions.

SCC test results and consequent times to failure are shown in table 10 for austenitic stainless steel samples tested in uninhibited and in inhibited brine solutions. Table 10 shows that the average time to failure for samples tested in uninhibited solution is approximately 119 hours. During the tests, many pits were noticed on the surface of this austenitic stainless steel samples. Further visual examinations have carried out after sample failure. This examination showed many small cracks on the surface and some of them were combined together to form bigger ones. Table 10 shows also that the average time to failure for samples tested in Molybdate inhibited solution is approximately 54 hours.

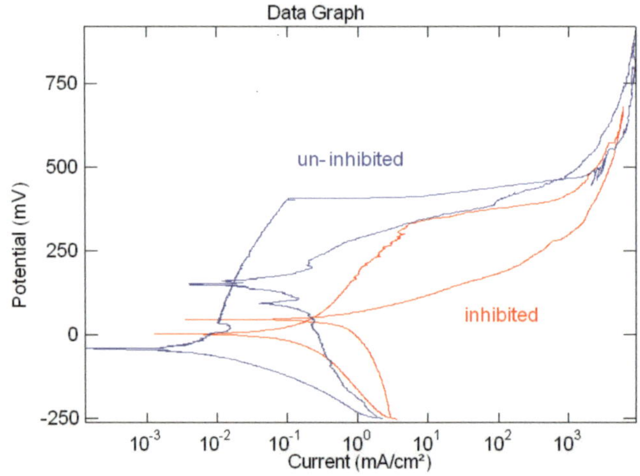

Figure 21. Cyclic sweep test plots for austenitic stainless steel in hot brine without inhibition and with inhibition. All measured potentials are measured versus austenitic stainless steel reference electrode [29].

Material	Austenitic stainless steel UNS S 31603 without inhibition			Austenitic stainless steel UNS S 31603 after inhibition		
	Sample 1	Sample 2	Sample 3	Sample 1	Sample 2	Sample 3
Time to failure (Hours)	160.29	76.38	119.56	32.56	58.53	71.59

Table 10. Times to failure for austenitic stainless steel samples tested under SCC conditions in uninhibited and in Molybdate inhibited brine solutions [29].

http://dx.doi.org/10.5772/52318

Figure 22 below is an attempt to use Evans's diagrams to possibly explain the behaviors for the metals under the inhibited and the uninhibited environments.

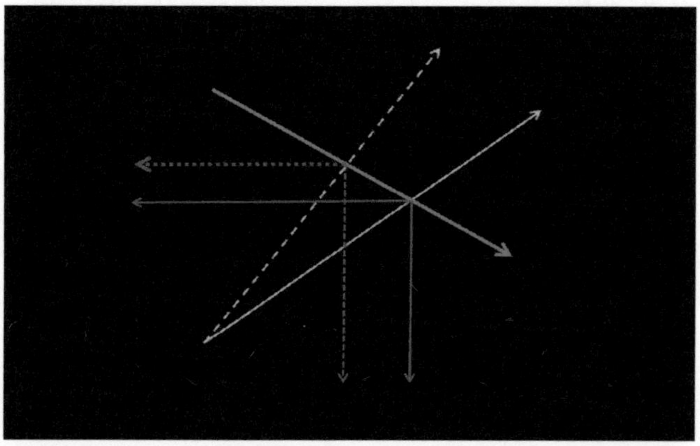

Figure 22. Evans Diagram illustrating results from cyclic sweeps due to environment chemical inhibition [29].

As per the diagram, inhibition of the environment using a passivating type, Molybdate inhibitor caused a relative noble shift in the ASS free corrosion potential. This change in the potential is possibly as a result of inhibitor molecules causing passivation on the metal surface. The shift to the more noble potential could be leading to a reduction in the uniform corrosion rate of the metal. However, this apparent reduction in uniform corrosion rate is not persistent and when the metal is stressed under real life operating conditions (simulated in cyclic sweeps in the lab tests) the intrinsic metal passivity might be adversely affected. The measured shift in potential and this apparent change in passivity are not necessarily enhancing metal resistance for localized corrosion (pitting). In fact this is adversely affecting the metal passivity as cyclic sweeps are showing a relative decrease to the pitting potential when compared to the uninhibited pitting potential. These results are also confirmed by the visual pitting observed on the metal after the test exposure and by the reduction in time to failure in the SCC tests.

6. Conclusions

1. Two common grades of ductile Ni-resist cast irons are widely used for sea water applications which include brine circulation pump casings. The first one is made as per ASTM A439 D2 (denoted in table 1 by D-material), whereas, the second one is made as per BS 3468 S2W (denoted by G-material), which has better weldability. The service lives of pump casings made of these two materials had been reported to be considerably different. The microstructure of each type of cast irons is different in terms of nodularity of graphite nodules, nodule count per square millimeter and uniformity of distribution of chromium carbides. This difference in microstructure is reflected in var-

iations, in hardness and tensile strength of both alloys. The D-type Ni-resist ductile iron has a relatively higher bulk hardness and higher tensile properties. However, electrochemical corrosion tests, in brine solution at room temperature, have shown similar corrosion behavior, in terms of corrosion rates, potential and polarization.

2. SCC tests in brine environment have indicated that D material has higher resistance to SCC than that of G material. SEM micrographs indicated that SCC failed specimens had distinctive fracture surface pattern identical to that of examined failed pump casings. This emphasizes that the cause of pump failure is SCC. The difference in behavior in resistance to SCC of the two materials is attributed to mechanical and microstructural properties. Further studies are recommended on the effects of carbides and nodules features, in DNI, on the resistance to SCC. This will allow optimization of SCC resistance of these materials.

3. Two types of stainless steel have been recommended as better substitutes to DNI in building brine circulation pump casings. These are austenitic UNS S31603 and superduplex UNS S32750. The corrosion resistance of the super duplex steel is relatively higher than that of the austenitic steel. This was demonstrated by a lower corrosion potential or enhanced passivity. Under the reported test conditions austenitic stainless steel showed clear breakdown in its passivity indicated by electrochemical cyclic sweeps, pitting potentials and visual observations.

4. Austenitic stainless steel showed susceptibility to SCC when loaded to 95% of its yield strength, polarized to a potential close to its pitting potential and exposed to fresh circulating brine at 55–60 °C. Pitting of austenitic stainless steel under these conditions is believed to stimulate crack initiation and hence the start of SCC. The lower strength and pitting resistance of austenitic stainless steel are believed to be the main reasons for its lower resistance to SCC compared to the super duplex steel.

5. Super duplex stainless steel showed immunity to SCC under the above mentioned testing conditions. Pitting is the mode of attack of most passive materials and can also be found on super duplex stainless steels possibly at higher corrosion accelerating potentials.

6. Fracture section of failed SCC austenitic stainless steel is characterized by having two zones; the first has step like features and facets analogous to cleavage facets, and the second corresponds to a dimpled fracture section, characteristic of ductile mechanical fracture.

7. Compared to austenitic stainless steel, super duplex stainless steel stands better chances of longer life as an engineering material used for building brine and sea water pump components.

8. Corrosion resistant alloys are expensive due to the presence of alloying elements that are essential to the corrosion resistance of these metals under challenging corrosive environments. Corrosion resistant alloys rely on their intrinsic passivity acquired from alloying additions.

9. Trying to enhance the passivity using Molybdate as passivating type inhibitors did not prove to help enhance corrosion resistance of ASS in hot brine environment. Under the given test conditions and using 350 ppm of Molybdate corrosion inhibitor, ASS showed localized corrosion in the form of pitting and failed in relatively shorter times in SCC tests. Thus using Molybdate as corrosion inhibitor does not eliminate the need for higher and more expensive alloy metallurgy to further improve corrosion resistance and SCC. However, still the idea of trying other inhibitor chemistries such as nitrites, carboxylates, orthophosphates, phosphates and other synergistic- types for inhibiting ASS in hot brine environment worth trying.

Author details

Osama Abuzeid[1], Mohamed Abou Zour[2], Ahmed Aljoboury[3] and Yahya Alzafin[4]

*Address all correspondence to: o.abuzeid@uaeu.ac.ae

1 Mechanical Engineering Department, UAE University, Al- Ain, The United Arab Emirates

2 General Electric Water & Process Technologies ME, Dubai, The United Arab Emirates

3 Industrial Support Services, Abu Dhabi, The United Arab Emirates

4 Dubai Electricity & Water Authority, Dubai, The United Arab Emirates

References

[1] http://en.wikipedia.org/wiki/Multi-stage_flash_distillation#cite_note-0. Retrieved 18-7-2012.

[2] A. I. Aljoboury, A.-H. I. Mourad, A. Alawar, M. Abou Zour and O.A. Abuzeid, "Stress corrosion cracking of stainless steels recommended for building brine recirculation pumps", Journal of Engineering Failure Analysis, Elsevier, 17,pp: 1337–1344, 2010.

[3] Private communication with pump manufacturer; 2007.

[4] ANSI/ NACE Standard TM0177, 1996.

[5] ASTM Standard G49-85, 2000.

[6] Y. A. Alzafin, A.-H. I. Mourad, M. Abou Zour, O. A. Abuzeid, "A Study on the failure of pump casings made of ductile Ni-resist irons used in desalination plants", Journal of Engineering Failure Analysis, Elsevier, Volume 14, Issue 7, pp 1294- 1300, 2007.

[7] Y. A. Alzafin, A.-H .I. Mourad, M. Abou Zour, O. A. Abuzeid,"Stress Corrosion Cracking of Ni-resist Ductile Iron Used In Manufacturing Brine Circulating Pumps of Desalination Plants", Journal of Engineering Failure Analysis, Elsevier, 16, pp: 733-739, 2009.

[8] Standard test methods for tension testing of metallic materials [Metric], E 8M-00b Metric, 2000.

[9] Standard practice for preparing, cleaning and evaluating corrosion test specimens, ASTM G1-03, 2003.

[10] Standard specifications for austenitic ductile iron castings, ASTM A 1999; 439–83.

[11] Standard test methods and definitions for mechanical testing of steel products. ASTM A370-02; 2002.

[12] Miyasaka M, Ogure N. Corrosion 1987;.43(10):582–8.

[13] Materials for saline water, desalination and oilfields brine pumps. A nickel development institute technical series No. 11004, 2nd ed., 1995.

[14] Properties and applications of Ni-resist and ductile Ni-resist alloys. A nickel development institute reference book series No. 11 018, Nickel Development Institute; 1998.

[15] Smart NG, Hitchman ML, Ansell RO, Fortune JD. A study of the electrochemical properties of Ni-resist in 3% sodium chloride solution. Corros Sci 1994;36(9):1473–89.

[16] Case histories in failure analysis. ASM; 1979.

[17] Singh Raman RK, Muddle BC. Eng Failure Anal 2004;11:199–206.

[18] British Standard Specification for Austenitic Cast Irons. BS 3468; 1986.

[19] Private communication from Bryan Johnson. CSS Technical Support, Sulzer Pumps, UK; 2007.

[20] ASTM A 240; 2006.

[21] Private communication with the Vice President – Plant 1 (D, E and G), Generation Division, Dubai Electricity and Water Authority, The United Arab Emirates; 2010.

[22] Mill Test Reports. Provided by Industeel, the material manufacturer; 2007.

[23] Metallurgical Test Report. Provided by North American Stainless, the material manufacturer; 2007.

[24] ASTM A370-07; 2007.

[25] ASTM G5-94 [reapproved 2004].

[26] Kwok CT, Fong SL, Cheng FT, Man HC. J Mater Process Technol 2006;176:168–78.

[27] Breslin CB, Chen C, Mansfeld F. Corros Sci 1997;39(6):1061–73.

[28] Fractography. ASM handbook; 1987.

[29] O. A. Abuzeid, A.I. Aljoboury, M. Abou Zour, "Effect of Corrosion Inhibition on the Stress Corrosion Cracking of UNS S31603 Austenitic Stainless Steel", Advanced Materials Research Vols. 476-478 (2012) pp 256-262.

[30] E. Flick, Corrosion Inhibitors :An Industrial Guide, second edition, New Jersey: Noyes Publications. (1993).

[31] R. Nishimura and Sundjono, Effect of Chromate and Molybdate on stress corrosion cracking of type 304 austenitic stainless steel in Hydrochloric acid solution, Corrosion. 56 (2000), 361-370.

[32] A. Devasenapathi and V. S. Raja, Effect of externally added Molybdate on repassivation and SCC behavior of AISI 304 SS in 1M HCl, Corrosion. 52 (1996), 243-249.

Chapter 6

A Comparative Study Between Different Corrosion Protection Layers

Adina- Elena Segneanu, Paula Sfirloaga, Ionel Balcu,
Nandina Vlatanescu and Ioan Grozescu

Additional information is available at the end of the chapter

http://dx.doi.org/10.5772/52985

1. Introduction

Corrosion is known as the destruction of materials due to interaction with corrosive environment. The destruction caused by metallic corrosion has become a serious problem in the world economy. Metallic corrosion is in most cases an electrochemical process occurring between a metal and its environment involving oxidation and reduction reactions.

There are two general classifications of corrosion which cover most of the specific forms. These are: direct chemical attack and electrochemical attack. In both types of corrosion the metal is converted into a metallic compound such as an oxide, hydroxide or sulphate. The corrosion process always involves two simultaneous changes: the metal that is attacked or oxidized suffers what may be called anodic change, and may be considered as undergoing cathodic change.

Direct Chemical Attack: Direct chemical attack, or pure chemical corrosion, is an attack resulting from a direct exposure of a bare surface to caustic liquid or gaseous agents. Unlike electrochemical attack where the anodic and cathodic changes may be taking place a measurable distance apart, the changes in direct chemical attack are occurring simultaneously at the same point.

The appearance of the corrosion varies with the metal. On aluminum alloys and magnesium it appears as surface pitting and etching, often combined with a gray or white powdery deposit. On copper and copper alloys the corrosion a greenish film, on steel reddish rust. When the gray, while, green, or reddish deposits are removed, each of the surfaces may appear etched and pitted, depending upon the length of exposure and severity of attack. If these surface pits are not too deep, they may not significantly alter the strength of the metal, however, the pits may become sites for crack development. Some types of corrosion can travel beneath surface coatings and can spread until the part fails.

Corrosion of steel or iron substrates can be slowed by coating the metal with different protective coatings. Porphyrins and multifunctional nanocomposites are examples of such coatings. Incorporation of organic or inorganic particles in the protective layers improves the physico-chemical properties [1].

Aluminum is an active metal and is naturally passivized forming an aluminum oxide film (Al_2O_3) on the metal surface. The oxide film can protect aluminum from corrosion in natural and some acid environments; however, it is expected to dissolve in alkalizes.

When oxygen is present (in the air, soil, or water), aluminum instantly reacts to form aluminum oxide. This aluminum oxide layer is chemically bound to the surface, and it seals the core aluminum from any further reaction. This is quite different from oxidation (corrosion) in steel, where rust puffs up and flakes off, constantly exposing new metal to corrosion.

Aluminium oxide is an amphoteric oxide (can react as either an acid or base) of aluminium with the chemical formula Al_2O_3. It is also commonly referred to as alumina, corundum, sapphire, ruby or aloxite in the mining, ceramic and materials science communities. It is produced by the Bayer process from bauxite. Its most significant use is in the production of aluminium metal, although it is also used as abrasive due to its hardness and as a refractory material due to its high melting point [2].

Components	[%]
Al_2O_3	99.2
SiO_2	0.06
Fe_2O_3	0,04
Na_2O	0.40
CaO	0.05

Table 1. Alumina characteristics

Alumina coatings are widely used in a range of industrial applications to improve corrosion protection, wear and erosion resistances, and thermal insulation of metallic surfaces. Refined alumina surfaces with long-term use are obtained from various efficient and adjustable processes. It can be seen that cost efficient arc-sprayed Al coatings post-treated by plasma-electrolytic oxidation (PEO) form Al_2O_3 -layers with remarkable corrosion protection, hardness, bonding strength, and abrasion resistance, as well as with the extended service time. The properties of these coatings are compared with alumina coatings obtained by flame spraying and atmospheric plasma spraying.

The type of corrosion observed on aluminum alloys are as follows:

a. *General Dissolution:* This occurs in strongly acidic or strongly alkaline solutions though there are specific exceptions. Certain inorganic salts (for example, aluminum ferric and zinc chlorides) hydrolyze in solution to give acidic or alkaline reaction and thus cause general dissolution. Lower alcohols and phenols in anhydrous condition do not allow protective layer to form and cause corrosion. Also at temperatures above 90ºC, the metal is uniformly attacked by aqueous systems.

b. *Pitting:* This is most commonly encountered form of aluminum corrosion. In certain, near neutral aqueous solutions, a pit once initiated will continue to propagate owing to the solution within the pit becoming acidic and the alumina not able to form a protective film close to the metal. Solutions containing the chlorides are very harmful in this respect particularly when they are associated with local galvanic cells, which can be formed for example by deposition of copper from solution or by particles such as iron unintentionally embedded in the metal surface. As little as 0.02 parts per million of copper in hard water could initiate pitting, although more is required for soft water. Aluminum is corroded by sea water. In alkaline media, pitting may occur at mechanical defects in the oxide. The aluminum alloys weather outdoors to grey color which deepens to black in industrial atmospheres.

c. *Intercrystalline Corrosion:*This is also electro thermal in nature, the galvanic cell being formed because of some heterogeneity in the alloy structure which may arise from certain alloying elements present.

d. *Stress Corrosion:*This form of corrosion is of limited occurrence with only aluminum alloys, in particular the higher strength materials such as the Al-Zn-Mg-Cu type and some of the Al-Mg wrought and cast alloys with higher magnesium content. The occurrence of stress corrosion increases in these alloys after specific low temperature heat treatments such as stove enameling.

e. *Bimetallic Corrosion*:Aluminum is anodic to many metals and when it is joined to them with a suitable electrolyte, the potential difference causes a current to flow and considerable corrosion can result. In some cases surface moisture on structures exposed to an aggressive atmosphere can give rise to galvanic corrosion. In practice, copper, brasses and bronzes in marine conditions cause most trouble. The danger from copper and its alloys is enhanced by the slight solubility of copper in many solutions and its subsequent redepositions on the aluminum to set up local active cells.

Contact with steel is comparatively less harmful. Stainless steels may increase attack on aluminum notably in sea water or marine atmospheres but the high electrical resistance of the two surface oxide films minimize bimetallic effects in less aggressive environments. With salts or heavy metals notably copper, silver and gold, the heavy metal deposits on to the aluminum subsequently causes serious bimetallic corrosion [3].

Pitting corrosion is the most commonly encountered form of localized corrosion of aluminum. This type of corrosion is difficult to detect because pits are small and often covered by corrosion products on the metal surface. The mechanism of pitting corrosion is so complicated that it is not completely understood even today. Pitting corrosion for aluminum occurs in media with a pH between 4 and 8. Generally, pitting corrosion consists of the initiation and the propagation stages. Figure 2 shows the mechanism of pitting corrosion of aluminum. In the initiation stage, chloride ions (Cl^-) or other aggressive anions such as bromide and iodide anions break down the passive film by absorbing on the aluminum surface covered oxide film resulting in formation of micro-cracks. These corrosive anions and corrosion products result in acidic solution within the rupture of the film due to hydrolysis which becomes an active corroding anode.

The pitting potential Epit is the critical potential indicating the relative resistance to pitting corrosion. Pitting occurs above the pitting potential Epit.

The chloride ions can penetrate into the oxide film causing film breakdown and dramatically decrease the pitting potential resulting in lower resistance of pitting corrosion.

After pitting occurs the pits will continue to develop as a self-propagating mechanism. Aluminum at the bottom of the pits will continue to be oxidized because the pits become the anode and produce Al^{3+}. The oxidation reaction at the pit's bottom is given by:

$$2Al \rightarrow 2Al^{3+} + 3e^{-} \tag{1}$$

The positive ions Al^{3+} will react with negative anions such as Cl^{-} resulting in acid chloride solution at the inside of pits. The acid pit solution can be concentrated up to a pH < 3 by further anodic dissolution. The aluminum dissolution process leads to self-propagating of pit growth because the concentrated acid pit solution becomes very corrosive.

Phosphogypsum is a waste product resulted from the process of obtaining the phosphoric acid from apathy and phosphorite by extraction with sulphuric acid.

$$Ca_5F(PO_4)_3 + 5H_2S0_4 + 10H_2O = 3H_3P0_4 + 5CaS0_4 \cdot 2H_20 + HF \tag{2}$$

This waste product is a very fine, wet and friable sand with characteristics dependant both on the origin of the ores as well as the various treatments applied to it. Regarding the crystalline structure, there are 4 types of phosphogyps: acicular crystals (80-500 mm); tubular crystals (40-200 mm); compact crystals; spherical polycrystallinic aggregations (50-100 mm).

Components [%]	**Origin:**	
CaO	33,31	32,68
SO_3	46,18	44,75
SiO_2	0,23	0,72
P_2O_5	0,84	1,16
H_2O	18,70	18,98

Table 2. The chemical composition of phosphogypsum

Phosphogypsum is used, as an additive, up to 10-15% compared to the binder (ash: lime) in the composition of light masonry blocks. Phosphogypsum can be used with soda as sulphate activator, for the mixed alkaline – sulphate activation of ash, in a mixture (ash: lime: phosphogypsum) for stabilizing foundation lands.

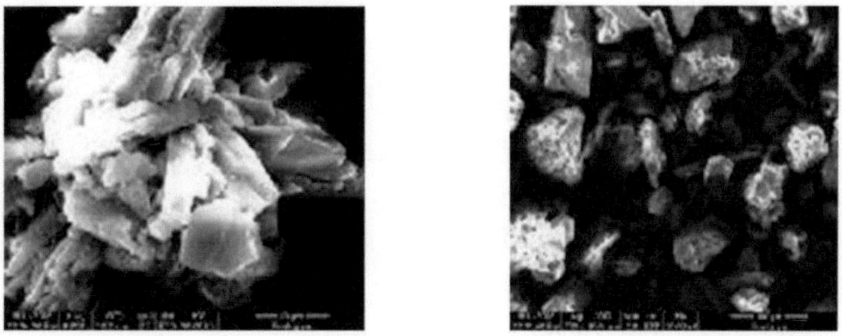

Figure 1. SEM micrographs of phosphogypsum films, 30 μm

Phosphogypsum is often used interchangeably in the research literature with the terms chemical, synthetic, waste or by-product gypsum, which is produced as solid chemical hazardous waste or by-product in industries, by wet or dry processes (sulphuric acidulation of phosphate rocks).

The disposal of phosphogypsum was very simple matter in the early days of industry, as plants had very low capacities (often producing only 25 tons per day of P_2O_5 versus a typical plant today is rated at one kilo tons per day and about up to five kilo tons per day) and environmental concerns and regulations were insignificant. Therefore, phosphogypsum is categorized as hazardous waste in Environmental Protection Agency (EPA) and under category 16 of hazardous waste (Management and Handling) rules 1989 framed by Government of India.

The proper utilization of phosphogypsum is imperious necessary to solve environmental and disposal problems. In different countries attempts have been made from time to time to find ways and means of utilizing phosphogypsum and therefore disposing significant quantity of waste. Concrete, which is a very important constituent of modern construction, is already adopted for waste management system, with exemplary applications like High Volume Fly ash (HVFA) technique. Indeed, concrete in particular is being used as a construction material of a number of waste disposal units, be they for operation, processing or storage of wastes [5,6,7,8].

The goal of this study is to develop of two different nanocomposite materials composed by porphyrin, a granular material (alumina or phosphogypsum) and alkyd paint:

i. porphyrin, a granular material (alumina) and paint (grey varnish paint and yellow alkyd paint);

ii. porphyrin and a granular material (phosphogypsum) and paint (grey varnish paint and yellow alkyd paint).

The anticorrosive properties of these multifunctional systems have been comparatively investigated on carbon-steel electrodes.

The porphyrin type was another parameter that was considered for determining effectiveness of the nanocomposite materials.

In our studies we used two types of modified porphyrins: Na_4TFPAc porphyrin ($C_{44}H_{26}N_4Na_4O_{12}S_4 \times H_2O$) and H_2TPP porphyrin. These porphyrins were dissolved in alkaline, acid and organic solutions (KOH, H_2SO_4 and benzonitrile) as follows:

1. 0.2 g of Na_4TFPAc porphyrin ($C_{44}H_{26}N_4Na_4O_{12}S_4 \times H_2O$) dissolved in 40 ml 10% KOH, mentioned from this point forward as system I;
2. 0.2 g of Na_4TFPAc porphyrin ($C_{44}H_{26}N_4Na_4O_{12}S_4 \times H_2O$) dissolved in 40 ml 10% H_2SO_4, mentioned from this point forward as system II;
3. 0.2 g of H_2TPP porphyrin (5, 10, 15, 20 tetrakis 4 phenyl-21H, 23H) dissolved in benzonitrile, mentioned from this point forward as system III.

The working electrode is the carbon-steel (FeC) electrode with a 0.13 cm^2 active surface; the counter electrode is made of platinum with a 0.31 cm^2 active surface and the reference electrode is the saturated calomel electrode (SCE). All these electrodes are connected to the potentiostat. As base electrolyte we used 20% Na_2SO_4.

The voltammetry measurements were carried out at electrochemical potentials ranging between -1,000 ÷ 2,500 mV and a sweep rate of 100mV/s. The electrode was treated with 5,10,15,20 tetrakis (4 phenyl)-21H, 23H porphyrin (H_2TPP) dissolved in 40 ml benzonitrile. The immersion time was 5 minutes.

Through corrosion tests we determined the resistance of both types of coatings as corrosion inhibitors by cyclic voltammetry and salt spray chamber.

2. Experimental

Electrochemical techniques are powerful tools to provide a better understanding of corrosion phenomena. Mansfeld et al have studied these electrochemical processes and developed a number of electrochemical methods to characterize the corrosion resistance of different metals and alloys [14-16].

Through cyclic voltammetry and Tafel tests we compared the porphyrin systems and determined the best one. Using this system we determined the resistance of alumina and phosphogyps as corrosion inhibitors through the salt spray corrosion test ASTM B 117 [9]. The apparatuses used were: the PGZ 402 Dynamic EIS Voltammetry potentiostat with VoltaMaster 4 software version 7.08 manufactured by Radiometer Copenhagen and the DCTC 600 salt spray chamber manufactured by Angelantoni Industrie (Figure 7).

In order to determine the resistance of alumina and phosphogyps as corrosion inhibitors, we used the DCTC 600 dry salt spray chamber and the ASTM B 117 method. The test lasted 14

http://dx.doi.org/10.5772/52985

days. The 5% NaCl solution was prepared using 1 kg of pure NaCl dissolved in 20 liters of distilled water.

Following the studies, we applied the improved method of dipping the electrode in paint before and after passing through the granular material (alumina or phosphogypsum), thus obtaining the multifunctional system: porphyrin (0.2 g H_2TPP/40 ml benzonitrile) + 2 ml paint + 0.9 ÷ 1.8 g alumina and porphyrin (0.2 g H_2TPP/40 ml benzonitrile) + 2 ml paint + 0.9 ÷ 1.8 g phosphogypsum.

We used two types of paint: grey varnish paint and yellow alkyd paint (named yellow paint).

Alkyd paint	
Aspect	homogeneous liquid, viscous
Density 23°C	0, 94 ± 0,05 g/ml
Flow time	55 -70s
Nonvolatile substances, 0,2- 0,3g, 105°C ;10 min	min. 50 %
Film characteristics :	
Glaze 60°, min.	82 %
Drying time 23 ±2°C, 50±5 % Relative humidity: - Drying time to touch -(TipB) -Drying time in depth - (TipD)	2 hours 5 hours
Liquids Resistance: water, detergent pH = 7, HCl 3%, ethanol 25%, mineral oil	good, no changes after 24 hours
Elasticity, min.	6mm

Table 3. Characteristics of alchidic paint

3. Results and Discussion

Evaluation of nanocomposite materials by cyclic voltammetry and Tafel tests for an immersion time 5 minutes

From the plots obtained through ciclic voltammetry we determined the peak current (i_{peak}) and peak potential (ε_{peak}) – for increasing and decreasing polarisation, respectively – as well as the passivation potential (ε_{pas}), passivation current (i_{pas}) and oxygen release potential (ε_{O2}). From the Tafel tests we determined the corrosion current (i_{cor}), polarisation resistance (Rp), the corrosion rate (v_{cor}) and the correlation coefficient (C).

The cyclic voltammograms and the Tafel tests are shown in figures 4, 5, 6 and 7.

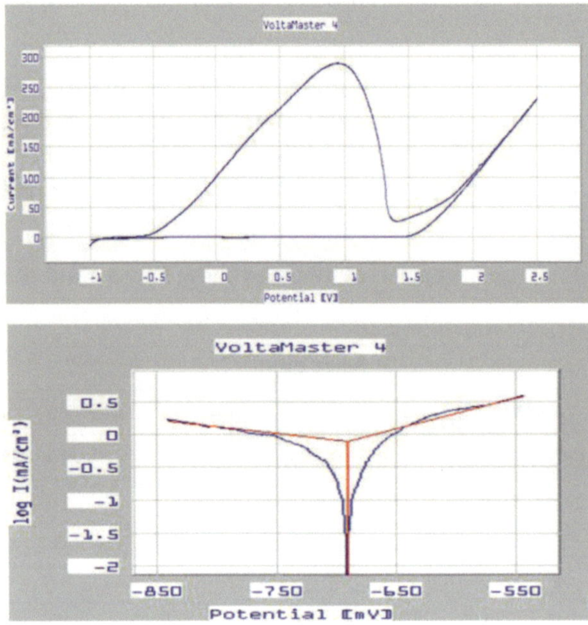

a)

b)

Figure 2. Uncoated carbon-steel electrode; 20% Na_2SO_4 support electrolite; polarisation speed – 100 mV/s; 25°C temperature. a – cyclic voltammogram; b – Tafel test

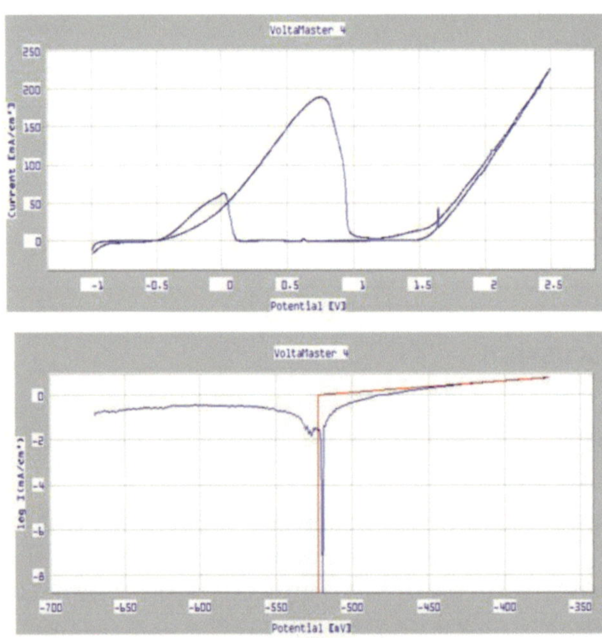

a)

b)

Figure 3. Coated FeC electrode; system I; a - cyclic voltammogram; b -Tafel test

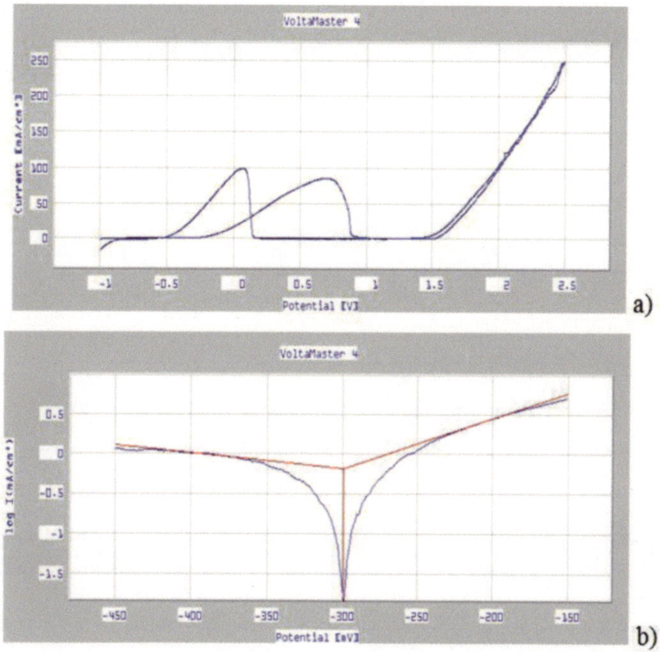

Figure 4. Coated FeC electrode; system II; a – cyclic voltammogram; b –Tafel test

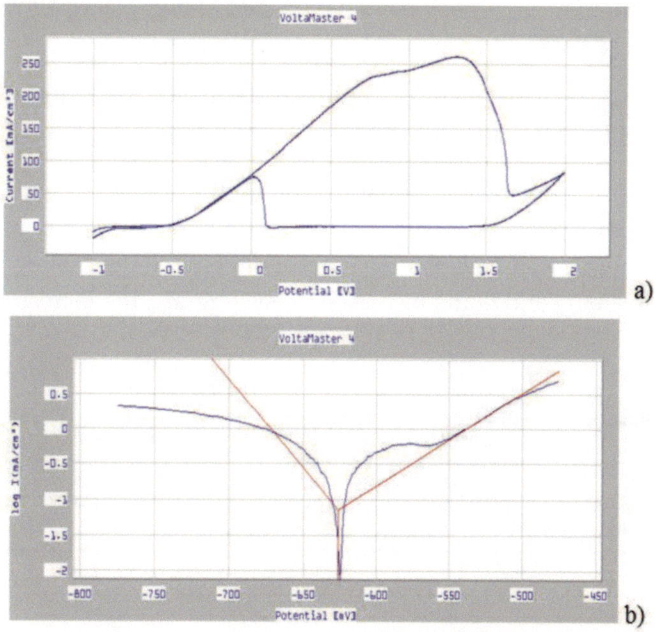

Figure 5. Coated FeC electrode according to system III; a – cyclic voltammogram; b –Tafel test

Table 4 shows the values of different parameters for uncoated and coated electrodes resulted from the cyclic voltammograms shown in figures 4a, 5a, 6a, 7a and table 3 shows the values of different parameters for uncoated and coated electrodes resulted from the Tafel tests shown in figures 4b, 5b, 6b, 7b.

Parameters	Electrodes			
	Uncoated	System I	System II	System III
$i^{\rightarrow}_{peak}$ [mA/cm^2]	290	280	90	80
$\varepsilon^{\rightarrow}_{pic}$ [mV]	900	750	600	1300
$i^{\leftarrow}_{peak}$ [mA/cm^2]	-	60	100	85
$\varepsilon^{\leftarrow}_{peak}$ [mV]	-	50	100	50
ε_{O2} [mV]	1500	1500	1500	1500
ε_{pas}[mV]	1350	900	850	1600
i_{pas} [mA/cm^2]	25	8	0	50

Table 4. Values of different parameters for uncoated and coated electrodes resulted from the cyclic voltammograms shown in figures 4a,5a, 6a, 7a

The notations used in table 4 are as follows: $i^{\rightarrow}_{peak}$ - peak current density for increasing polarization; $\varepsilon^{\rightarrow}_{peak}$ - peak potential for increasing polarization; $i^{\leftarrow}_{peak}$ - peak current density for decreasing polarization; $\varepsilon^{\leftarrow}_{peak}$ - peak potential for decreasing polarization; ε_{O2} - oxygen release potential; ε_{pas} - passivation potential; i_{pas} - passivation current.

Variation of $i^{\leftarrow}_{peak}$ is correlated with the porphyrin systems. Na_4TFPAc both in NaOH as well as in H_2SO_4 and H_2TPP solutions exhibits a decrease in $i^{\leftarrow}_{peak}$ value.

The peak potential ($\varepsilon^{\leftarrow}_{peak}$) shows a similar evolution: decreased values for system I and II and increase values for system III. There is no variation in the oxygen release potential (ε_{O2}). Passivation potential (ε_{pas}) and passivation current depend on the coating systems. The values decreased for system I and II and increased for system III (20% increase).

Parameters	Electrodes			
	Uncoated	System I	System II	System III
i_{cor} [mA/cm2]	0.9792	0.7666	0.6506	0.0718
v_{cor} [mm/year]	11.48	8.99	7.63	0.842
Rp	50.91	129.51	59.57	47.67
C	0.9962	0.9996	0.9997	1.000

Table 5. Values of different parameters for uncoated and coated electrodes resulted from the Tafel tests shown in figures 4b,5b, 6b, 7b.

The notations used in table 5 are as follows: i_{cor} - corrosion current; v_{cor} - corrosion rate; Rp - polarisation resistance; C - correlation coefficient.

Making a comparison between the treated and untreated electrodes observed a significant decrease in corrosion rate for the treated electrodes resulting in the formation of a passive layer of corrosion protection, also resulting in better adhesion for application of corrosion inhibitor on the surface electrode.

These results indicate that corrosion rate depends on the type of porphyrin system. For system I, there is a corrosion rate increase of about 25%; a corrosion rate decrease of 15 % for system II and a significant corrosion rate decrease of about 10 times for system III, indicating that the surfaces of FeC electrode coated with H_2TPP form an efficient passive corrosion protection layer. The reference corrosion speed is the v_{cor} of the uncoated FeC electrode (8.99 mm/year).

Electrochemistry as a tool for the study of the physical properties of porphyrins has been extensively used recently. The effect of substituents on the oxidation and reduction reactions of the π system of tetraphenylporphyrin (TPP) was investigated by Kadish et al. by cyclic voltammetry in methylene chloride and in other solvent. In all solvents it was found that electron-donating substituents shift the oxidation and the reduction potentials to more positive values. Electron attracting substituents on the other hand, shift the potentials to more negative values. When the $E_{1/2}$ values were plotted vs. the Hammett substituent constant σ a linear free-energy relationship were obtained. Both ring reduction to yield the π anion radical and ring oxidation to yield the π cation radical showed sensitivity to para substitution, and an average value of 0.07 ± 0.01 V was found for the Hammett reaction constant ϱ. The Hammett constants for the oxidation reactions to yield π cation radicals were found relatively more sensitive to para substitution and varied from 0.064 V for free-base TPP (in CH_2C1_2).

The electrode processes (oxidation and reduction) did not satisfy all the diagnostic criteria for reversible charge transfer: [9]

$$(p-X,p-Y)H_2TPP - e^- \leftrightarrow \left[(p-X,p-Y)\ H_2TPP\right]^+ \tag{3}$$

$$(p-X,p-Y)H_2TPP - e^- \leftrightarrow \left[(p-X,p-Y)\ H_2TPP\right]^+ \tag{4}$$

$$(p-X,p-Y)\ H_2TPP + e^- \leftrightarrow \left[(p-X,p-Y)\ H_2TPP\right]^- \tag{5}$$

$$\left[\ (p-X,p-Y)\ H_2TPP\right]^- + e^- \leftrightarrow [p-X,p-Y)\ H_2TPP]^{2-} \tag{6}$$

Porphyrins are a class of natural pigments containing a fundamental skeleton of four pyrrole nuclei united through the α-positions by four methine groups to form a macrocyclic structure. Porphyrin is designated also with the nomenclature of porphine. The common meso-substituted porphyrins are tetraphenyl porphyrin (R = phenyl) and ortho, meta or para substituted phenyl porphyrins. Usually, for the mesotetraphenylporphyrin synthesis is used pyrole and an aldehyde such as benzaldehyde, salicylaldehyde, and so on [12].

The porphyrin nucleus is a tetradentate ligand. When coordination occurs, two protons are removed from the pyrrole nitrogen atoms, leaving two negative charges. The porphyrin ring system exhibits aromatic character, containing 22 π-electrons, but only 18 of them are delocalized according to the Hückel's rule of aromaticity (4n+2 delocalized π-electrons, where n=4).

Simple porphyrins with identical substituents in *meso* or β-positions are usually prepared by methods based on monopyrrole condensation (the Rothemund Method), when four identical pyrrole molecules are condensed into a porphyrin in one step.

4. Redox Properties of Tetraphenylporphyrin (H_2TPP)

The use of electrochemical methods to estimate the redox properties of porphyrins is vital for understanding the photochemistry of porphyrins. In general, free-base porphyrins possess two oxidation peaks and two reduction peaks in cyclic voltammetry.

These correspond to the one and two electron oxidation and reduction of the porphyrin π system. The redox properties exhibit a good correlation with the electronegativity or inductive parameter of the central metal atom. Substituents on the porphyrin ring show a good correlation between the redox potentials and the Hamett σ values. The electrochemical band gap corresponds well with the optical band gap determined by the lowest energy absorption in the Q band, indicating that the central metal and substituents equally affect the HOMO and LUMO levels. The number of substituents is also correlated with the shifts in the redox peak positions. Distortion from planarity seems to cause a dramatic change in the oxidation potential. The addition of a redox active metal complicates the overall electrochemical properties of porphyrins, due to the intervening oxidation and reduction potentials of the metal. The change in axial ligand also seems to play an important role in the redox potentials of porphyrins.

Electrochemistry provides valuable insight into the electronic properties of molecules. This technique provides information on the position of the energy levels, in particular the highest occupied molecular orbital (HOMO) and the lowest unoccupied molecular orbital (LUMO) are easily discernable from these measurements. The position of the HOMO of a molecule is probed by determining its anodic potential, while the position of the LUMO is determined by its cathodic potential. These positions can be referenced with respect to the vacuum level by adding 4.7 eV to the onset of the peak (oxidation/reduction) with respect to the ferrocene / ferrocenium (Fc / Fc+) redox couple. The redox properties of porphyrin were performed studies on different types of electrodes [13].

The electrochemical properties of tetraphenylporphyrin are well known. The first oxidation with respect to Fc/Fc+ was determined to be 0.54 V and was determined to be reversible. The

http://dx.doi.org/10.5772/52985

first reduction peak was determined to be located at -1.75 V and was also reversible. These observations correspond well with previously published results. The electrochemical "band-gap" (HOMO – LUMO gap), which was determined by the difference of the $E_{1/2}$ of the anodic and cathodic waves, was determined to be ~ 2.2 eV which corresponds well with the optical gap measured by absorption. Substitution on the *meso*-phenyl groups of TPP has little effect on the charge transport properties of these complexes. Therefore these novel complexes still facilitate whole transport and hinder electron transport. In order to correct this problem, the electron transporting ability must be increased. This can be accomplished by either lowering the LUMO by substitution of electron withdrawing moieties on the pyrroles of the porphyrin or by creating a "molecular wire" in which electrons can flow freely, reducing the barrier for electron injection [10].

In experiments, the efficiency of nanocomposites materials developed as inhibitors for corrosion was tested in salt spray chamber. The corrosion of metallic surface occurs in the aggressive chloride ions attack in the oxide film and the corrosion are propagated according to the following anodic reactions:

$$Al \rightarrow Al^{+3} + 3e \tag{7}$$

$$Al^{+3} + 3H_2O\ Al(OH)_3 + 3H^+ \tag{8}$$

Hydrogen evolution and oxygen reduction are the important reduction processes at the intermetallic cathodes as:

$$2H^+ + 2e \rightarrow H2 \tag{9}$$

$$O_2 + 2H_2O + 4e \rightarrow 4OH \tag{10}$$

According to reaction 2, the pH will decrease as corrosion propagates. To balance the positive charge produced by reactions 1 and 2, chloride ions will migrate into the pit. The resulting HCl formation inside the pit causes in the protective film accelerated the corrosion propagation. It is postulated that, at the critical pitting potential, breakdown occurs by a process of field assisted Cl adsorption on the hydrated oxide surface and formation of a soluble basic chloride salt which readily goes into solution.

5. Evaluation of coatings by exposure to the salt spray test

The table below shows the conditions for testing the anticorrosive coating in salt spray chamber [11].

Type of saline used	NaCl and distilatted water
pH	6,5-7,5
Concentration of the solution	5 %
Spray pressure	60-150 kPa
The amount of saline spray	1-2 ml/h per 80 cm^2
Temperature	25- 40°C
Sample pozition	15° vertical
Time	336 hours

Table 6. Conditions inside the salt spray chamber for testing the anticorrosive coating

In Table 7 are presented the corrosion evolution based on the composition of the protecting film.

PROTECTION FILM COMPOSITION	CORROSION EVOLUTION (h)
H_2TPP + yellow paint + 1,8g Al_2O_3	After 23h: pitting corrosion
H_2TPP + yellow paint + 0,9g Al_2O_3	After 23h: pitting corrosion
H_2TPP+grey varnish paint + 0,9g Al_2O_3	After 48h: uniform corrosion
H_2TPP + grey varnish paint + 1,8g Al_2O_3	After 23h: pitting corrosion
grey varnish paint + 0,9g Al_2O_3	After 25h: pitting corrosion
grey varnish paint + 1,8g Al_2O_3	After 48h: uniform corrosion
yellow paint + 0,9g Al_2O_3	After 23h: pitting corrosion
grey varnish paint + 1,8 g Al_2O_3	After 77h: uniform corrosion
untreated	After 1h: uniform corrosion
H_2TPP + yellow paint + 1,8g phosphogyps	After 23h: pitting corrosion
H_2TPP + yellow paint + 0,9g phosphogyps	After 48h; pitting corrosion
H_2TPP + grey varnish paint + 1,8g phosphogyps	After 144h: uniform corrosion
H_2TPP + grey varnish paint + 0,9g phosphogyps	After 5h; corossion was present in two places in the form of blisters
grey varnish paint + 0,9g phosphogyps	After 48h; uniform corrosion
grey varnish paint + 1,8g phosphogyps	After 77h; uniform corrosion
grey varnish paint + 0,9g phosphogyps	After 23h; uniform corrosion
grey varnish paint + 1,8 g phosphogyps	After 23h; uniform corrosion

Table 7. Corrosion evolution based on the composition of the protecting film

6. Conclusions

- Corrosion speed is mostly decreased in case of the electrode treated with porphyrin, gray varnish paint and 1.8 g phosphogyps.

- The method used for covering the electrodes is also important because when the electrode is diped in paint its surface becomes uniformly covered thus avoiding exogen corrosion.

- The multifunctional system containing phosphogypsum was a better corrosion inhibitor than the one containing alumina.

Author details

Adina- Elena Segneanu*, Paula Sfirloaga, Ionel Balcu, Nandina Vlatanescu and Ioan Grozescu

*Address all correspondence to: s_adinaelena@yahoo.com

National Institute of Research and Development for Electrochemistry and Condensed Matter, INCEMC-Timisoara, Romania

References

[1] Saji, V. S., & Thomas, J. (2007). Nanomaterials for corrosion control. *Current Science*, 92(1), 1.

[2] Harris, C., & Mc Lachlan, R. (1998). Clark, Colin- Micro reform- impacts on films: aluminium case study. Melbourne: Industry Commission

[3] Mahindru, D. V. (2011). Ms Priyanka Mahendru, Protective Treatment of Aluminum and its Alloys. Global Journal of Research in Engineering Version 1.0,, 11(3)

[4] Huang, Y. (2009). Electrochemical Evaluation Of Advanced Anodized Aluminum and Chromate-Free Conversion Coatings. *A Dissertation Presented to the Faculty Of The Graduate School University Of Southern California.*

[5] Environmental Protection Agency. (1999). CFR Part 61: National emission standard for hazardous air pollutants. *National Emission standards for Radon emissions from phosphogypsum stacks, Federal register*, 64(2), 5574-5580.

[6] Jamwal, N. (2000). Blind to danger: Special report. *Down to Earth*, 8(17).

[7] Thakare, R. B., Bhatia, O. P., & Hiraskar, K. G. (2001). Phosphogypsum utilisation in India: Literature survey. *The Indian Concrete Journal*, http://www.icjonline.com, 75(6), 408-410.

[8] Singh, M. (1980). Physio- chemical studies on phosphogypsum for use in building materials. *PhD Thesis, University of Roorkee, Roorkee, India.*

[9] Ulman, A., Manassen, J., Frolow, F., & Rabinovich, D. (1981). Synthesis and Properties of Tetraphenylporphyrin Molecules Containing Heteroatoms Other Than Nitrogen. *Electrochemical Studies Inorg. Chem.*, 20.

[10] Senge, M. O. (2000). *In "The Porphyrin Handbook"; Kadish, K. M., Smith, K. M., Guilard, R., Eds., Academic: New York*, 1, 239-347.

[11] Standard practice for operating salt spray (fog) apparatus. B117-02.

[12] Sharma, R. K., Ahuja, G., & Sidhwani, I. T. (2009). A new one pot and solvent-free synthesis of nickel porphyrin complex. Green Chemistry Letters and Reviews June, , 2(2), 101-105.

[13] Kadish, K. M., & Caemelbecke, E. V. (2003). *J. Solid. State Electrochem*, 7, 254.

[14] Mansfeld, F. (1976). The Polarization Resistance Technique for Measuring Corrosion Currents. In: Fontana MG, Staehle RW, editors. Advances in Corrosion Science and Technology New York: Plenum Press , 6, 163.

[15] Mansfeld, F., & Kendig, M. W. (1984). *Corrosion*, 41, 490.

[16] Mansfeld, F., Han, L. T., Lee, C. C., & Zhang, G. (1998). Evaluation of corrosion protection by polymer coatings using electrochemical impedance spectroscopy and noise analysis. *Electrochimica Acta*, 43, 2933.

Section 3

Corrosion of Biomaterials

Chapter 7

Galvanic Corrosion Behavior of Dental Alloys

Hamoon Zohdi, Mohammad Emami and
Hamid Reza Shahverdi

Additional information is available at the end of the chapter

http://dx.doi.org/10.5772/52319

1. Introduction

There is a wide variety of dental alloys, ranging from nearly pure gold and conventional gold-based alloys to alloys based on silver, palladium, nickel, cobalt, iron, titanium, tin, and other metals (Table 1). The types of dental alloys have increased significantly since 1980s in order to change the market price of gold and palladium. Although gold alloys are the materials of choice in this area because of their high mechanical properties, good corrosion resistance and excellent biocompatibility, their price still poses the essential challenge to dentistry. So that, alternative materials such as Ag-Pd alloys, Co-Cr alloys and Ti alloys have been introduced into dentistry [1,2].

Alloy type	Uses in dentistry	Major elements
Gold-based	Restorations, solders	Au, Ag, Cu, In, Pd, Pt, Zn
Palladium-based	Restorations	Pd, Ag, Ga, Cu
Silver-based	Restorations, solders	Ag, Pd
Cobalt-based	Restorations	Co, Cr, Mo, Fe, C, Si, Mn
Nickel-based	Restorations, orthodontic materials	Ni, Mo, Fe, C, Be, Mn
Titanium-based	Implants	Ti, O, N, C, Fe, H
Iron-based	Implants, orthodontic materials	Fe, C, Ni, Cr
Mercury-based	Amalgam	Hg, Ag, Sn, Pd, Cu, In

Table 1. Common types of alloys in dentistry and their major component elements [1]

Dental alloys can be classified into a variety of applications such as restorations, amalgam, implants, solders and orthodontic materials. The used alloys should have suitable physical, mechanical and chemical properties for mentioned applications. For example, an orthodontic wire is required to have a relatively high flexibility (a low modulus) and the ability to be bent and shaped. However, the alloy for a dental restoration should have almost no flexibility (a high modulus) and be hard and difficult to deform.

Biocompatibility is an important measuring property which should be evaluated first. The word biocompatibility is defined as the ability of a material to perform with an appropriate host response in a specific situation [3]. Although the biological compatibility of dental alloys inclined to be considered separately from the other properties, biocompatibility is rather related to other properties of the alloys such as corrosion resistance which is estimated by measuring the release of the corrosion products themselves. The higher the corrosion rate of the alloy, the greater the metal ion release and the greater the risk of undesirable reactions in the mouth. These reactions may include unpleasant metallic tastes, allergy, irritation or another reaction. Since the release of metal ions depends on electrochemical rules, many efforts have been made to evaluate the biocompatibility of dental alloys via electrochemical analyses [1, 4, 5].

As was mentioned earlier, the corrosion behavior of dental materials is important because poor biocompatibility of the products may render the materials inappropriate for implantation. In general, the word corrosion stands for material or metal deterioration or surface damage in an aggressive environment. The oral environment is also favorable for corrosion in which the metal is attacked by presence of natural agents (air and water), temperature fluctuations (hot and cold meals) and pH changes because of diet (milk products or orange juice), resulting in partial or complete dissolution, deterioration, or weakening of any solid substance [4, 6]

One of the problems associated with the use of metallic materials in dentistry is the probability of galvanic corrosion [2, 4, 7-10]. Generally, galvanic corrosion is either a chemical or an electrochemical corrosion. This phenomenon is attributed to a potential difference between two different metals connected through a circuit for current flow to occur from more active metal (more negative potential) to the more noble metal (more positive potential). In addition, galvanic corrosion is a very complex phenomenon. Six basic factors are involved in galvanic corrosion: (1) potentials, (2) polarization, (3) electrode areas, (4) resistance and galvanic current, (5) the electrolyte medium, (6) aeration, diffusion and agitation of the electrolyte [11]. Galvanic coupling is a galvanic cell in which the more negative metal (anode) is the less corrosion resistant metal than the more positive metal (cathode) [12]. The resulting galvanic couple achieves a mixed potential that reaches between the corrosion potentials of the uncoupled metals (Fig. 1). Due to mutual polarization, the anodic corrosion rate of the anode will be accelerated, while the anodic rate of the cathode will be reduced [10].

In dentistry application, galvanic corrosion occurs when two or more dental prosthetic devices with dissimilar alloys come into contact while subjected to oral liquids like salvia; the difference between the corrosion potentials results in a flow of electric current between them. Therefore, the galvanic cell is formed and causes the increasing corrosion rate of the anode and enhancing the amount of ion metal released. The galvanic current passes not only through the metal/metal connections, but also through the tissues, which may cause pain. Galvanic currents in the oral environment may cause sharp pain when they exceed 20 mA [13]. Geis-Gerstorfer et al. [14] believes that the galvanic corrosion of dental devices is important in two respects: 1) the biological effects which may result from the dissolution of alloys and 2) the current flow resulting from galvanic cell that could cause bone destruction. The galvanic corrosion may be started due to the interaction of prosthetic devices. For example, a restoration or prosthesis in physical contact with amalgam in an adjacent tooth or between dental implants, fillings or crowns [9].

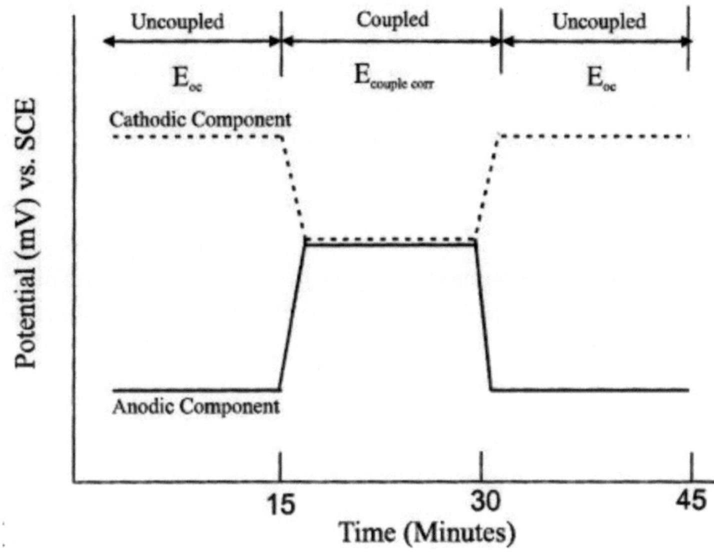

Figure 1. Schematic of data acquired during continuous potential measurement [8].

2. Methods

The measurement of the biocompatibility of dental alloys is a complicated issue. However, tests for biocompatibility assessment are classified as either in vitro or in vivo tests. In vitro tests are performed outside a living organism, while in vivo tests are conducted in an animal's body. In vitro tests are the cheapest and fastest of the biocompatibility tests, but because they are not performed in a living system, their significance is often subjected. Conversely, in vivo tests are more informative than in vitro due to the fact that the device is subjected to all dimensions of the biological response, but they are also expensive and highly complex to control and interpret.

In addition, biocompatibility is relatively related to other properties of the alloys such as corrosion having a direct relationship with the release of metal ions. The presence of metal ions in the body may cause various phenomena such as transportation, metabolism, allergy, carcinoma and accumulation in organs. Therefore, measuring the metal ion release of biomaterials (dental alloys) is important as well as other biocompatibility tests, which is done by methods like atomic absorption spectroscopy, inductively coupled plasma mass spectroscopy, or X-ray fluorescence spectroscopy.

Moreover, since the release of metal ions depends on electrochemical rules, many efforts made to evaluate the biocompatibility of dental alloys by corrosion tests in in vitro and in vivo studies. For this specific case, as was discussed earlier, galvanic corrosion can enhance the corrosion rate of the anode resulting in high amount of metal ion released. Zero Resistance Ammetry is the main method used to evaluate galvanic corrosion behavior of dental alloys; with ZRA probes, two electrodes of dissimilar metals are exposed to the process fluid. When immersed in solution, a natural voltage (potential) difference exits between the electrodes. The current generated due to this potential difference relates to the rate of corrosion which is occurring on the more active of the electrode couple. A schematic of the experimental setup is shown in Fig. 2. Besides, the measurement of currents and potentials in galvanic couple or uncoupled electrodes has been made to obtain more information. Moreover, the electrochemical corrosion tests like open circuit potential, cyclic and linear polarization, potentiostatic polarization or electrochemical impedance spectroscopy (EIS) have been developed for many years to estimate the degree of corrosion on dental alloys by measuring the current flow during the corrosion process, or change in potential of the alloy relative to some standard.

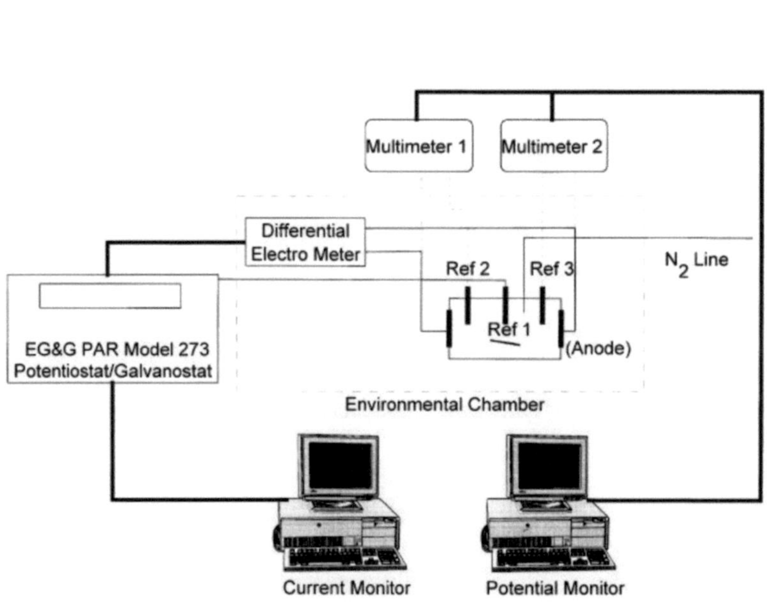

Figure 2. Schematic diagram for the galvanic cell set-up [8].

3. *In vitro* and *in vivo* tests

The aim of this section is to evaluate and compare, in vitro and in vivo, the galvanic corrosion behavior of dental alloys such as restorations, amalgam, implants or orthodontic materials when they are used in a mouth at the same time. It was indicated in the previous section that the simultaneous using of these devices can cause some biologic problems due to the galvanic corrosion effect.

Nowadays, titanium and titanium alloys are widely used in odontology because of their excellent characteristics such as good mechanical behavior, low density, high corrosion resistance in body fluids and excellent biocompatibility. The high biocompatibility of these alloys is attributed to the formation of the passive film (TiO_2) on the surface, which is highly protective. As these alloy implants and prosthetic devices become more common, the galvanic interaction with other metallic materials may become an issue [9, 11, 15, 16]. Studies of galvanic cells of titanium with dental alloys indicated either almost no interaction, or very small galvanic currents [11, 15, 17]. R. Venugopalan and L. C. Lucas [8] used continuous corrosion potential monitoring in coincidence with Zero Resistance Ammetry to achieve galvanic corrosion properties of restorative and implant materials coupled with titanium. All tests were carried out in artificial salvia solution. The composition of the electrolyte is shown in Table 2. They found that noble restorative (Au-, Ag-, and Pd-based) alloys coupled to titanium are least susceptible to galvanic corrosion, while the Ni–Cr–Be alloy showed unstable galvanic corrosion behavior. Also findings of N. M. Taher and A. S. Al Jabab [2] indicated that the highest galvanic corrosion resistant alloys coupled with titanium implant abutment material were Pontallor (Au-based), Ternary Ti, R800 (Co-Cr alloy) and Jelstar (Ag-Pd alloy), respectively. But, RCS (Ni–Cr) alloy was found to be highly susceptible to galvanic corrosion, which is in accordance with former study. In general, it should be mentioned that titanium was anodic to noble alloys and cathodic to iron and nickel-based passivating alloys. It is also worth noting that other researchers could obtain relatively similar results [11, 18].

Compound	Composition (g/dm^3)
K_2HPO_4	0.20
KCl	1.20
KSCN	0.33
Na_2HPO_4	0.26
NaCl	0.70
$NaHCO_3$	1.50
Urea	1.50
Lactic acid	Up to pH = 6.7

Table 2. Chemical composition of the artificial salvia

Dental amalgams are still the most common metallic direct filling materials. However, because of mercury toxicity, low-mercury and mercury-free gallium-based direct filling alloys have been developed in recent years. Both dental amalgams and gallium-based filling alloys are passivating materials with less protective passive film in comparison with titanium or most other passivating dental alloys. Researchers [9, 19] have considered the galvanic interaction between titanium and direct filling alloys like gallium and variety of copper containing amalgams. They reported that the galvanic interaction between titanium and direct filling alloys is small. The gallium alloy was the most sensitive to galvanic corrosion among the samples when in contact with titanium, which is ascribed to relatively poor corrosion resistance of gallium alloys [20]. It was shown that the galvanic corrosion resistance of mentioned alloys coupled to Ti from the highest to lowest are as follows: High copper dental amalgam > Low copper dental amalgam > Gallium-based direct filling.

Nitinol (Nickel titanium) is a very attractive material for use as an orthodontic wire due to its unique shape memory and superelasticity properties. Researchers evaluated the galvanic corrosion of these orthodontic wires with dental alloys in artificial saliva. They found that placing stainless steel brackets or Aristaloy (Ag-Cu-Sn) amalgam in direct contact to nitinol arch wire is not recommended, because it causes enhanced corrosion rate of nitinol arch wire. They also suggested that using ceramic brackets instead of stainless steel brackets could help to get rid of the occurrence of galvanic corrosion [21].

Another group of metals used in dentistry is chromium based alloys. Ciszewski et al. [4] investigated the galvanic corrosion behavior of Remanium GM 380 (chromium–cobalt alloy) and Remanium CS (chromium–nickel alloy) when bound together or coupled with Amalcap plus (silver-based amalgam) in an artificial saliva solution at 37°C. It was found that a bimetallic cell consisting of Remanium CS and Remanium GM 380 alloys has a very low EMF (electromotive force) which is not a potential source of galvanic currents in the oral cavity. Conversely, galvanic cells prepared from Amalcap plus and Remanium CS or Remanium GM 380 showed a much greater EMF. This obviously showed that in these latter it is possible to expect some metal ions in the saliva solution as a result of galvanic currents. They also indicated that even elements from a cathode specimen of a galvanic cell are able to dissolve into the solution. These results, from an electrochemical point of view, are surprising.

Since in vivo tests are generally expensive, time consuming, controversial and complex to study, there are few reports on the galvanic corrosion behavior of dental devices in a living organism. Most researchers have performed an indirect measuring technique to determine in vivo galvanic currents of dental alloys [22-24]. Palaghias et al. [25] investigated in vivo behavior of gold-plated stainless steel titanium dental retention pins and showed that in vivo corrosion resistance of the titanium pins was superior to that of gold-plated stainless steel pins. Besides, Nilner et al. [26] found that gold-gold couples have lower galvanic currents than those of amalgam-amalgam and amalgam-gold couples; they indicated that galvanic currents for the couples are generally below 15 mA, which is below the threshold of pain (20 mA) [13]. It should be mentioned that the galvanic corrosion behavior of dental alloys is expected to fluctuate over time due to various factors, including changes in the pH and composition of saliva, disruption of the alloy's passive film due to chewing, and aging of the restoration. Changes may also take place because of thermal and mechanical stresses [10, 27, 28]. As a result, it is concluded that the interpretation of this type of test is so complicated and needs more time to investigate.

4. Parameters affect galvanic corrosion

One of the important factors affecting the galvanic corrosion is the surface area ratio of the two dissimilar alloys (cathode/anode). An unfavorable area ratio, which consists of a large cathode and small anode, may cause a higher corrosion [12]. Reports showed that the galvanic potential and current density increased with the increasing Ti/alloy area ratio, where Ti plays the role of cathode. Therefore, the higher galvanic corrosion occurred [9, 29]. Besides, reducing the surface area of the anode by 75 percent increases the galvanic activity of stainless steel/nitinol couple [30]. However, Iijima et al. found that the different anodic/cathodic area ratios (1:1, 1:2.35, and 1:3.64) had little effect on galvanic corrosion behavior of stainless steels and titanium bracket alloys coupled with four common wire alloys: nickel-titanium alloy, β-titanium alloy, stainless steel and cobalt-chromium-nickel alloy [31].

Fluoride is well known as an effective caries prophylactic agent and its systemic application has been recommended widely over recent decades to be the main method for preventing plaque formation and dental caries. Toothpastes, mouthwashes, and prophylactic gels contain from 200 to 20,000 ppm F(-) and can impair the corrosion resistance of dental alloys in the oral cavity [32, 33]. Anwar et al. [34] considered the effect of fluoride ion concentration on the corrosion behavior of Ti and Ti6Al4V implant alloys, when coupled with either metal/ceramic or all-ceramic superstructures. It was shown that increased fluoride concentration leads to a decrease in the corrosion resistance of all tested couples. Moreover, findings of Johansson and Bergman showed that adding fluoride to the solution made the titanium potential more active and enhanced the corrosion of titanium in combination with high-copper amalgams [29]. In fact, authors have exhibited that increase in concentration of NaF (fluoride ion) decreases the corrosion resistance of NiTi arch wires [35, 36] and titanium implants in different solutions [37, 38]. It is also demonstrated that the combination of low pH and presence of fluoride ions in solution severely affect the breakdown of the protective passivation layer that normally exists on nitinol and titanium alloys, leading to pitting corrosion [39, 40, 41].

Another parameter which could affect galvanic current is the initiation of localized corrosion (pitting and crevice corrosion). Mastication and other food contents (such as chloride ions) may initiate localized corrosion of dental alloys. This type of corrosion once initiated the corrosion current density and therefore the galvanic current increase. The presence of pitting on Nitinol arch wire harshly increases the galvanic corrosion rate of the anode, which indicates that dentists and researchers should be aware of other types of corrosion as well as galvanic one to investigate dental alloys appropriately. It was also mentioned that, initiation of localized corrosion on anode increased the galvanic current by up to 45 times revealing that consideration of the effect of localized corrosion on galvanic corrosion is necessary [21].

5. Preventing galvanic corrosion

According to the published literature and experimental results, the set-up for an acceptable couple combination in the mouth environment could be defined as the following: (1) the difference in $E_{o.c.}$ of the two materials and the $I_{couplecorr}$ should be as small as possible; (2) the $E_{couplecorr}$ of the couple combination should be significantly lower than the breakdown potential of the anodic component and (3) the repassivation properties of the anodic component of the couple should also be acceptable [8]. Besides, the use of some special composites, ceramics and metallic glasses can improve galvanic corrosion behavior of dental alloys. Metallic glasses, called also glassy or amorphous metals are rapidly quenched alloys explained as a metastable class of materials with no long range periodic lattice structure. These alloys are considered to be the materials of future [42, 43]. J.-J. Oak et al. [44, 45] developed new Ti-based bulk metallic glassy (BMG) alloys for application as biomaterials. Ti-based amorphous alloys containing no harmful elements (Ni, Al, Be) are expected to exhibit high potential for dental materials. The Ti-based amorphous ribbons exhibited good bend ductility, higher strength and lower Young's modulus than pure Ti and Ti–6Al–4V alloy. In addition, Ti-based amorphous alloys had an excellent potentiality of corrosion resistance that were passivated in wide passive range and at the lower passive current density in simulated body fluid conditions. It is demonstrated that the $Ti_{44.1}Zr_{9.8}Pd_{9.8}Cu_{30.38}Sn_{3.92}Nb_2$ bulk glassy alloy has a high potentiality to be applied in dental implant devices. These materials can be applied as coatings on the amalgams or restorative alloys to improve corrosion resistance of substrates.

6. Summary

In this review, we have highlighted comprehensive study of galvanic corrosion behavior of dental alloys. The titanium /titanium alloys, gold, silver-palladium and cobalt-chromium are main classes of alloys widely used as dental implants. In general, although they have exceptional properties which make them ideal for corrosion and wear resistance dental applications, it has been reported that failures of some implants are due to the galvanic-type corrosion. The galvanic current passes through the metal/metal junctions, which may finally cause pain owing to release of metal ions. The oral environment is particularly favorable for corrosion. The corrosive process is mainly of an electrochemical nature and natural saliva presents a good electrolyte. Fluctuations in temperature (hot and cold meals), changes in pH because of diet (milk products or acid dressings), and decomposition of food all contribute to the process. It is also mentioned that the parameters like the surface area ratio of the two dissimilar alloys, pH and the presence of fluoride could severely affect galvanic corrosion. To measure galvanic corrosion, researchers have investigated direct coupling or galvanic experiments which are conducted on restorative and implant materials coupled to another dental device like amalgam. They launched a comparative assessment of the electrochemical measurements attained using different methods and different preparation to study this type of corrosion. Zero Resistance Ammetry is the main method used to evaluate galvanic corrosion behavior of dental alloys in in vitro and in vivo. Besides, in this review, it is shown that new types of prosthesis/implants like metallic glasses (ribbons) could be applied as new generation of implants with excellent corrosion properties. These materials can be applied as coatings on the dental alloys to improve corrosion resistance of substrates.

Author details

Hamoon Zohdi, Mohammad Emami and Hamid Reza Shahverdi*

*Address all correspondence to: shahverdi@modares.ac.ir

Department of Materials Engineering, Tarbiat Modares University, Tehran, Iran

References

[1] R. Messer and J. Wataha, Encyclopedia of materials: science and technology, Elsevier Science Ltd, Oxford (2002), ISBN: 0-08-043152-6, pp. 1–10.

[2] N. M. Taher, A. S Al Jabab, Galvanic corrosion behavior of implant suprastructure dental alloys, Dent. Mater. 19 (2003) 54-59.

[3] J. Black, Biological Performance of Materials: Fundamentals of Biocompatibility, Marcel Dekker, New York (1992).

[4] A. Ciszewski, M. Baraniak, M. Urbanek-Brychczyn´ska, Corrosion by galvanic coupling between amalgam anddifferent chromium-based alloys, Dent. Mater. 23 (2007) 1256–1261.

[5] T. Hanawa, Metal ion release from metal implants, Mater. Sci. Eng. C 24 (2004) 745-752.

[6] D.Upadhyay, M. A. Panchal, R.S. Dubey, V.K. Srivastava, Corrosion of alloys used in dentistry: A review, Mater. Sci. Eng. A 432 (2006) 1–11.

[7] Y.Fovet, J. Pourreyron, Y. Gal, Corrosion by galvanic coupling between carbon fiber posts and different alloys, Dent. Mater. 16 (2000) 364–373.

[8] R. Venugopalan, L. C. Lucas, Evaluation of restorative and implant alloys galvanically coupled to titanium, Dent. Mater. 14 (1998) 165–172.

[9] N. Horasawa, S. Takahashi, M. Marek, Galvanic interaction between titanium and gallium alloy or dental amalgam, Dent. Mater. 15 (1999) 318–322.

[10] E. J. Sutow, W. A. Maillet, J. C. Taylor, G.C. Hall, In vivo galvaniccurrents of intermittently contacting dental amalgam and other metallic restorations, Dent. Mater. 20 (2004) 823–831.

[11] B. Grosgogeat, L. Reclaru, M. Lissac, F. Dalard, Measurement and evaluation of galvanic corrosion between titanium/Ti6Al4V implants and dental alloys by electrochemical techniques and auger spectrometry, Biomaterials 20 (1999) 933-941.

[12] N. Peres, Electrochemistry and corrosion science, Kluwer Academic Publishers, Boston, 2004.

[13] J. Karov, I. Hinberg, Galvanic corrosion of selected dental alloys, J. Oral Rehabil. 28 (2001) 212-219.

[14] J. Geis-Gerstorfer, J. G. Weber, K. H. Sauer, In vitro substance loss due to galvanic corrosion in Ti implant/Ni-Cr supraconstruction systems, Int. J. Oral Max. Impl. 4 (1989) 119-123.

[15] L. Reclaru and J. M. Meyer, Study of corrosion between a titanium implant and dental alloys, J. Dent. 22 (1994) 159-168.

[16] G. R. Parr, L. K. Gardner, R. W. Toth, Titanium: the mystery metal of implant dentistry, Dental materials aspects, J. Prosthet. Dent. 54 (1985) 410-414.

[17] G. Ravnholt, Corrosion current and pH rise around titanium coupled to dental alloys, Scand. J. Dent. Res. 96 (1988) 466–472.

[18] M. Cortada, Ll. Giner, S. Costa, F. J. Gil, D. Rodri´guez, J. A. Planell, Galvanic corrosion behavior of titanium implants coupled to dental alloys, J. Mater. Sci. - Mater. Med. 11 (2000) 287-293.

[19] J.D. Bumgardner, B.I. Johansson, Galvanic corrosion and cytotoxic effects of amalgam and gallium alloys coupled to titanium, Eur. J. Oral Sci. 104 (1996) 300–308.

[20] N. Horasawa, S. Takahashi, Corrosion resistance of gallium alloys, J. Jpn. Dent. Mater. 3 (1996) 192–201.

[21] A. Afshar, M. Shirazi, M. Rahman, E. Fakheri, Effect of localized corrosion on the galvanic corrosion of nitinol and dental alloys, Anti-Corros. Method M. 56 (2009) 323 – 329.

[22] M. Bergman, O. Ginstrup, K. Nilner, Potential and polarization measurements in vivo of oral galvanism, Scand. J. Dent. Res. 86 (1978) 135-145.

[23] M. Bergman, O. Ginstrup, B. Nilsson, Potentials of and currents between dental metallic restorations, Scand. J. Dent. Res. 90 (1982) 404-408.

[24] B. Johansson, E. Stenman, M. Bergman, Clinical study of patients referred for investigation regarding so-called oral galvanism. Scand. J. Dent. Res. 92 (1984) 469-475.

[25] G. Palaghias, G. Eliades, G. Vougiouklakis, In vivo corrosion behavior of gold-plated versus titanium dental retention pins, J. Prosthet. Dent. 67 (1992) 194–198.

[26] K. Nilner, P-O. Glantz, B. Zoger, On intraoral potential and polarization measurements of metallic restorations. Acta Odontol. Scand. 40 (1982) 275-281.

[27] B. Johansson, L. Lundmark, Direct and indirect registration of currents between dental metallic restorations in the oral cavity, Scand. J. Dent. Res. 92 (1984) 476-479.

[28] E. J. Sutow, W. A. Maillet, G. C. Hall, Corrosion potential variation of aged dental amalgam restorations over time, Dent. Mater. 22 (2006) 325–329.

[29] B.I. Johansson, B. Bergman, Corrosion of titanium and amalgam couples: Effect of fluoride, area size, surface preparation and fabrication procedures, Dent. Mater. 11 (1995) 41–46.

[30] J.A. Platt, A. Guzman, A. Zuccari, , D.W. Thornburg, B.F. Rhodes, Y. Oshida, B. Moore, Corrosion behavior of 2205 duplex stainless steel, Am. J. Orthod. Dentofacial Orthop. 112 (1997) 69-79.

[31] M. Iijima, K. Endo, T. Yuasa, H. Ohno, K. Hayashi, M. Kakizaki, I. Mizoguchi, Galvanic corrosion behavior of orthodontic archwire alloys coupled to bracket alloys, Angle Orthod. 76 (2006) 705-711.

[32] J. Noguti, F. de Oliveira, R. C. Peres, A. C. M. Renno, D. A. Ribeiro, The role of fluoride on the process of titanium corrosion in oral cavity, Biometals 25 (2012) 859-862.

[33] F. Rosalbino, S. Delsante, G. Borzone, G. Scavino, Influence of noble metals alloying additions on the corrosion behaviour of titanium in a fluoride-containing environment, J. Mater. Sci. - Mater. Med. 23 (2012) 1129–1137.

[34] E. M. Anwar, L. S. Kheiralla, R. H. Tammam, Effect of fluoride on the corrosion behavior of Ti and Ti6Al4V dental implants coupled with different superstructures, J. Oral Implantol. 37 (2011) 309–317.

[35] H. Benyahia, M. Ebntouhami, I. Forsal, F. Zaoui, E. Aalloula, Corrosion resistance of NiTi in fluoride and acid environments, Int. Orthod. 7 (2009) 322–334.

[36] T. H. Lee, T. K. Huang, S. Y. Lin, L. K. Chen, M. Y. Chou, H. H. Huang, Corrosion resistance of different nickel-titanium archwires in acidic fluoride-containing artificial saliva, Angle Orthod. 80 (2010) 547–553.

[37] W. Cheng, H. Yu, X. Lin, X. Han, L. Liu, T. Ding, P. Ji, The effect of fluoride on dental alloys in different concentrations of fluoride artificial saliva. Hua Xi Kou Qiang Yi Xue Za Zhi 30 (2012) 18–21.

[38] S. Kumar, T.S.N. Sankara Narayanan, S. Saravana Kumar, Influence of fluoride ion on the electrochemical behaviour of b-Ti alloy for dental implant application, Corros. Sci. 52 (2010) 1721–1727.

[39] N. Schiff, B. Grosgogeat, M. Lissac, F. Dalard, Influence of fluoride content and pH on the corrosion resistance of titanium and its alloys, Biomaterials 23 (2002) 1995–2002.

[40] B. Lindholm-Sethson, B.I. Ardlin, Effects of pH and fluoride concentration on the corrosion of titanium, J. Biomed. Mater. Res. 86A (2008) 149–159.

[41] B. G. Liang, X. T. Shen, L. Liu, Y. X. Lu, Z. D. Yu, C. X. Yang, Y. Z. Zhang, Effect of pH value and fluoride ions on corrosion resistance of pure Ti and Ni–Cr–Ti alloy in artificial saliva, Zhejiang Da Xue Xue Bao Yi Xue Ban 39 (2010) 399–403.

[42] M. Miller, P. Liaw (Eds.), Bulk Metallic Glasses, Springer, New York, 2008.

[43] H. Zohdi, H. R. Shahverdi, S. M. M. Hadavi, Effect of Nb addition on corrosion behavior of Fe-based metallic glasses in Ringer's solution for biomedical applications, Electrochem. Commun. 13 (2011) 840-843.

[44] J.-J. Oak, D. V. Louzguine-Luzgin, A. Inoue, Investigation of glass-forming ability, deformation and corrosion behavior of Ni-freeTi-based BMG alloys designed for application as dental implants, Mater. Sci. Eng. C 29 (2009) 322–327.

[45] J.-J. Oak, A. Inoue, Formation, mechanical properties and corrosion resistance of Ti-Pd base glassy alloys, J. Non-Cryst. Solids 354 (2008) 1828-1832